- • DATE			
JUN 03 1996			
JUL 29 1996			
NOV 23 1996			
DEC 16 1996			
JAN 27 1997			
MAY 20 1997			
SEP 09 2000			
OCT - 7 2003			
NOV 23 2005			
DEC 04 2010			
APR 07 2015			

Planets

A Smithsonian Guide

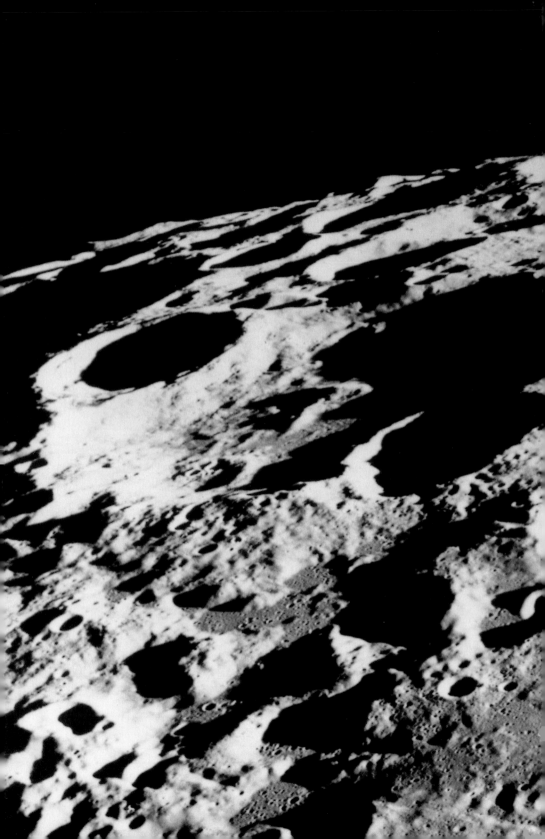

Planets
A Smithsonian Guide

Thomas R. Watters
Chairman of the Center for Earth and
Planetary Studies
National Air and Space Museum

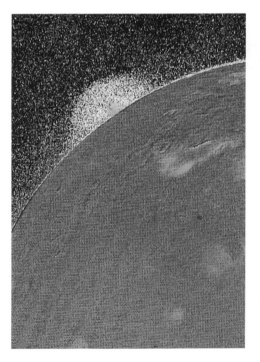

Macmillan • USA

Macmillan • USA
A Simon & Schuster Macmillan Company
15 Columbus Circle
New York, NY 10023

A Latimer Book

In the Gallery

page 1. This composite image shows the gas giant Jupiter with its four major satellites.

pages 2–3. Breathtaking detail in the rings of Saturn was captured by *Voyager 2's* cameras in 1981.

pages 4–5. The *Apollo 11* mission provided this dramatic image of highlands on the far side of the Moon.

pages 6–7. Jupiter's satellite, Io, stands out against turbulent clouds on the planet's southern hemisphere.

page 8. Scattered sunlight within the upper atmosphere of Titan, Saturn's largest moon, creates a beautiful, blue crescent.

page 9. A volcanic plume erupts from the surface of Io.

page 11. This fiery view of the Sun shows hundreds of bright granules surrounded by dark areas. Each granule is about the size of Texas.

Library of Congress Cataloging-in-Publication Data

Watters, Thomas R.
  Planets: Smithsonian Guides / Thomas R. Watters
    p.   cm.
  "A Latimer book."
  Includes bibliographic references and index.
  ISBN 0-02-860404-0   ISBN 0-02-860405-9 (pbk.)
  1. Planets [1. Planets]. I. National Air and Space Museum.
  II. Title. III. Series.
QB602.W38   1995
523.4—dc20   94-44718 CIP AC

Author's Dedication
To my wife Nancy and my children, James, Samantha, and Adam.

Ligature Inc.

Publisher	**Production**	**Series Design**	**Editorial**
Jonathan P. Latimer	Anne E. Spencer	Patricia A. Eynon	Susan Judge
	Paul Farwell		Mary Ashford
Contributing Writer		**Design**	Elizabeth Grube
Richard A. Bolster		Lisa Rosowsky	Toni J. Rosenberg
		Julia Sedykh	

The Smithsonian Institution   **Macmillan**

Trish Graboske   Mary Ann Lynch
Jim Wilson   Laura C. Wood
                                                  Robin Besofsky

Manufactured in Hong Kong
10 9 8 7 6 5 4 3 2 1

MAY 14 '96

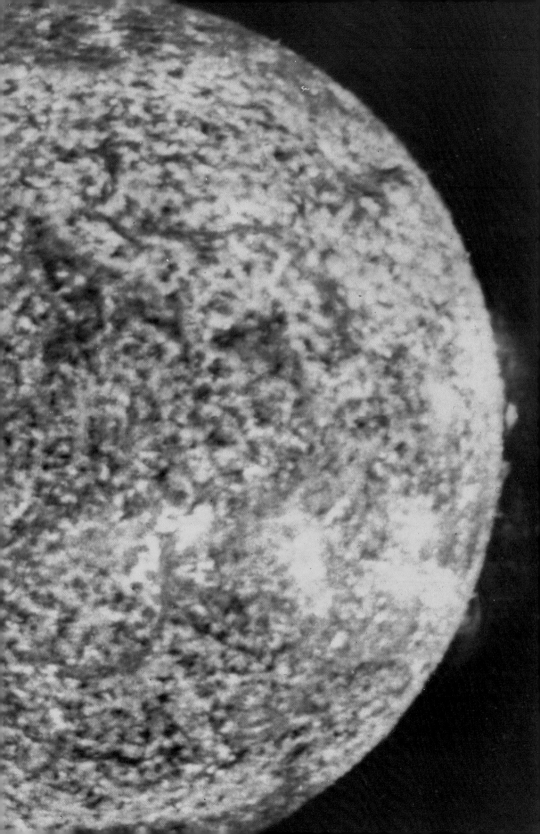

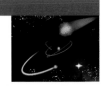

Unveiling the Solar System

Throughout history people have been fascinated by the planets—by what they are, how they move across the sky, and how these other worlds relate to Earth. Our concept of the solar system has changed dramatically over the centuries, although new views often were not welcomed. Now that the age of planetary exploration has begun, we have seen and learned things our ancestors never imagined. But while our knowledge has increased tremendously, our sense of wonder about the planets has not diminished.

The fifteenth-century Mongol prince, Ulug Bek, studied the sky from this observatory in the city of Samarkand in central Asia. The 130-foot (40 m) marble arc supports a sliding platform used to record the positions of stars and planets.

Early Views

For thousands of years ancient peoples observed the Sun, the Moon, the visible planets, and the stars. Drawing on their religions and cultural traditions, they often explained what they saw in the heavens by weaving myths and filling the sky with both angry and benevolent gods as well as with places sacred to those gods. To the Egyptians, for example, the swath of white light that cuts across the black summer night sky—now known as the Milky Way galaxy—was the heavenly Nile River in the land of the dead.

Ancient peoples also discovered practical applications based upon their observations. They noted the position of the Sun in the various seasons and its effect on crop growth. They saw that the Moon affected tides. And they saw that the stars appeared to remain fixed in relation to one another, while the planets moved through the stars. Realizing this, they used the stars as points of reference to guide their travels.

Written records, observatories, monuments, measuring instruments, and other artifacts show that most ancient civilizations not only studied the skies, but did so with considerable accuracy. The Sumerians, Babylonians, Egyptians, and Chinese made the first known recorded measurements of the stars and the movements of the planets.

By about the fifth century B.C., the Babylonians had identified the region of the sky through which the Sun, Moon, and planets move as viewed from Earth, and they had named several constellations. Among their other contributions, the Babylonians developed a calendar based on the movements of the seven "planet gods" visible to the human eye—the Sun, Moon, Mercury, Venus, Mars, Jupiter, and Saturn (the basis of our names for the seven days of the week).

Records of the Han dynasty, which lasted from 202 B.C. to 220 A.D., show that Chinese astronomers observed everything from eclipses and planetary movements to comets and sunspots.

The Ancient Observatory in Beijing, China, was built in 1296. The Chinese led the world in astronomy during the medieval period. They noted two supernovas—exploding stars that become extremely bright for several years and then fade—in 1006 and 1054. These observations helped astronomers in the twentieth century to understand how supernovas happen.

Studies of an ancient African observatory in Kenya strongly suggest that an accurate and complex astronomy-based calendar system was developed there by the first millennium B.C. In the Americas, manuscripts of the Mayans reveal a wealth of astronomical observations including details of lunar and solar eclipses and a Venus calendar. And long before the study of astronomy was established in Europe, it flourished in the Arabian empire, which stretched from Spain to India and lasted from the mid-eighth to the fifteenth century. In fact, many of the measuring instruments that Arab astronomers used were superior to those used by the Greeks.

Greek Contributions

The word *planet* comes from the Greek word meaning "wanderers." The Greeks named the region of the sky through which the planets pass the *zodiac* and divided it into twelve groups of stars, or constellations. The word *zodiac* means "circle of animals," and most of the constellations were given names of animals. The Greeks were also noted for the application of precise geometric measurements to astronomy that led to amazing accuracy in descriptions of planetary movements. For example, Eratosthenes (276–194 B.C.), librarian at the museum in Alexandria, Egypt, used geometric calculations related to the position of the Sun and concluded that the circumference of Earth was 25,000 miles (40,000 km), which is very nearly correct. Other Greeks, including Aristarchus and Hipparchus, also made key contributions.

During the 300s B.C., the Greek philosopher Aristotle incorporated the geocentric (Earth-centered) model of the universe into his teachings. This picture (above) of the geocentric model is from *Cosmographia*, published centuries later in 1539.

This drawing (above) of the general early conception of the universe is from a book published in 1614.

Aristarchus In about 270 B.C., Aristarchus of Samos proposed the *heliocentric* model of the solar system, which correctly placed the Sun at the center with Earth and the other planets revolving around it. He reached his conclusion after estimating the relative sizes of Earth, the Moon, and the Sun and discovering that the Sun was much larger than Earth. But the idea was disregarded until nearly 1,900 years later, largely because of the strength of the prevailing *geocentric* belief, based on Greek philosophy and religion, that Earth itself was the center of the solar system.

Hipparchus Perhaps the greatest of the astronomers of antiquity was Hipparchus, who died sometime after 127 B.C. He developed the first star catalog, listing some 850 stars, and he calculated the length of a year to within 6.5 minutes. He also calculated the *precession*, or wobble, of the Earth's axis by comparing his own observations with those made in Alexandria 150 years earlier and in ancient Babylonia. Like others of his time, however, Hipparchus rejected the heliocentric theory of Aristarchus.

The man in the left foreground of this colored woodcut created by French astronomer Camille Flammarion in the 1800s represents human curiosity about what exists beyond the observed heavens.

The Ptolemaic System

Observations of the planets showed that although they move generally toward the east through the constellations, they appear to turn back west at various intervals before resuming their eastward paths. This peculiar backward looping is now called *retrograde* motion. One explanation of this was proposed by Claudius Ptolemaeus (127–145 A.D.) of Alexandria, generally referred to as Ptolemy. Like most Greek astronomers, Ptolemy believed that celestial bodies moved in circles, which were considered the most perfect geometric shape. He suggested that the planets themselves, while orbiting around centers like wheels, also orbit the Earth in circles. This idea of "epicycles" provided an approximation to the retrograde behavior of the planets.

Basing his studies on the work of Hipparchus, Ptolemy expanded his predecessor's catalog to include 1,022 stars. Like Hipparchus, Ptolemy agreed with the geocentric view of the planetary system.

A Thousand Years of Silence

After the fall of the Roman Empire in 476 A.D., the important work of the Greek astronomers was mostly lost, forgotten, or ignored. In Europe the Ptolemaic model of the planetary system survived because it supported the philosophy of Aristotle and fit well with the teachings of the powerful Roman Catholic church. This belief that Earth was the center of the universe would remain unchallenged for more than 1,000 years in the Western world.

Nicolaus Copernicus, considered the founder of modern astronomy, dared to reject the Earth-centered Ptolemaic model. He could not prove his theory, but later astronomers like Galileo and Kepler would provide the evidence to support the heliocentric model.

Early European Astronomers

The Renaissance in Europe, which began in the 1300s, fostered a renewed interest in the ideas and culture of ancient Greece. In this stimulating intellectual atmosphere, some began to rethink the Ptolemaic model of the universe.

The Copernican Revolution

The Polish astronomer Nicolaus Copernicus, born in 1473, used mathematical proofs to revive the heliocentric model of Aristarchus. As part of this work, he launched an extensive series of observations and calculations to refine Ptolemy's tables. Copernicus claimed that all the planets, including Earth, orbit the Sun on perfectly circular paths. But he recognized that Earth moves rapidly through space and believed that this motion accounted for the apparent motion of the stars and even the retrograde motion of the planets.

Although many resisted the heliocentric theory, astronomers generally accepted Copernicus' highly detailed data. In his great book *On the Revolutions of the Heavenly Spheres,* published in 1543 only hours before he died, Copernicus showed that the motions of other celestial bodies could be explained by the motions of the Earth. This principle became the basis for some of the most important astronomical discoveries of the next two centuries.

A page from Copernicus' groundbreaking treatise *On the Revolutions of the Heavenly Spheres* contains his diagram of the solar system with the Sun at the center and six planets tracing perfect circles around it. Uranus, Neptune, and Pluto had yet to be identified.

Tycho Brahe

On August 21, 1560, a 14-year-old Danish boy named Tycho Brahe observed a total eclipse of the Sun, occurring just as existing almanacs had predicted. Three years later he observed that Jupiter and Saturn were aligned. This time the almanacs and the Copernican tables were days off in their predictions, so the young Tycho decided to produce more accurate tables. In 1572 he discovered a new star, or *nova*. Now believed to have been a *supernova*, a star that expels much of its mass, it disappeared from sight after 16 months. This discovery conflicted with the prevailing belief that the stars that made up the celestial sphere were fixed and could not change.

Tycho and his assistants continued to amass data, making amazingly accurate observations without the use of a telescope and correcting nearly every known astronomical table. By measuring a comet and showing that it was beyond the Moon, Tycho disproved the belief that comets were hot gases in Earth's upper atmosphere. As to the heliocentric theory of Copernicus, Tycho agreed that the planets revolved around the Sun but claimed that the Sun itself revolved around the fixed Earth.

Decades before the telescope was invented, Danish astronomer Tycho Brahe, depicted in a 1598 drawing, made accurate tables of the positions of celestial objects.

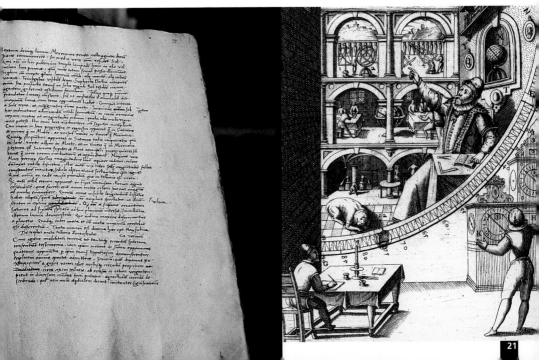

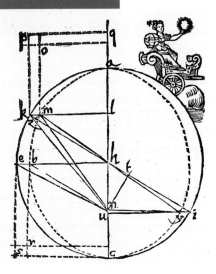

This illustration by Johannes Kepler appeared in his *Astronomia Nova* in 1609. It demonstrated the elliptical orbit of Mars, changing the belief that had existed for more than 2,000 years that orbits must be circular.

Tycho Brahe's brilliant student Johannes Kepler openly supported the Copernican model of the solar system. Kepler was also interested in optics and helped explain how the human eye functions.

Johannes Kepler

Born in a small German town in 1571, Johannes Kepler received his education through a scholarship program for the sons of poor families. He attended the University of Tübingen, where his astronomy teacher believed in the Copernican theory. When Tycho Brahe, impressed by a paper Kepler had written, invited the young student to join his research group, Kepler's genius began to flourish. The following year, in 1601, Tycho died, and Kepler was appointed to carry on his work, using Tycho's extraordinarily accurate data. By 1609 Kepler had announced two crucially important findings. In 1618 he announced a third.

First he declared that the orbits of the planets around the Sun are elliptical, rather than perfectly circular. Using Tycho's data about Mars and trying to fit various curves to its orbit, Kepler realized in 1609 that the orbit of Mars was an ellipse, not a circle or a combination of circles as had been thought. Once he made this discovery, he could explain the orbits of the other planets as ellipses as well.

Kepler's second major finding in 1609 was also based on his observations of Mars. He concluded that a planet moves faster when closer to the Sun and slower when farther from the Sun, sweeping out equal areas in space in equal periods of time.

Kepler's third discovery was the existence of a mathematical relationship between a planet's distance from the Sun and its orbital *period*—the length of time it takes to orbit the Sun once.

These three proven observations came to be known as Kepler's laws of planetary motion. They would finally replace the old established beliefs and unquestionably support the findings of Copernicus. They would also prove critical later as a foundation for Isaac Newton's theory of gravitational force.

Galileo

In Italy in 1609, Galileo Galilei built his first telescope without ever having seen one assembled. This telescope had two lenses and magnified objects to three times their size. He continued to study the heavens using telescopes he made by grinding his own lenses. The largest had a lens with a 1.7-inch (4.3-cm) diameter and a magnification factor of 33. In 1610 his pioneering work led to the discovery of four satellites of Jupiter. He also observed sunspots and craters on the Moon.

Galileo's observation that sunspots moved from day to day and thus must be on or close to the Sun's surface clashed with the Aristotelian concept of the perfection of the Sun. Also contrary to the prevailing view, Galileo agreed with Copernicus that Earth was not the center of the planetary system. His discovery that Venus has phases just like Earth's Moon confirmed that Venus orbits the Sun and supported the heliocentric model. But learned people of the time still could not accept this reality. When invited by Galileo to look through his telescope and judge for themselves, professors of philosophy at the University of Padua refused.

Early in the 1500s Martin Luther had challenged the authority of the pope, and by Galileo's time the Protestant Reformation was under way. Europe was being torn by a series of religious wars that had begun in 1618. The Roman Catholic church felt further threatened by scientific studies whose conclusions conflicted with its own teachings. Thus, church authorities forced Galileo to renounce his own theories as errors. His books were banned, he was forbidden to teach the Copernican view, and he was confined to his home from 1633 until his death in 1642.

Galileo (above), built his first telescope in 1609. Already skeptical of the theories of Aristotle and Ptolemy, he found his suspicions confirmed when he used this new tool to explore the sky. In the painting (left), Galileo is depicted inviting the chief magistrate of Venice to look through one of his telescopes.

Isaac Newton

Before the time of Isaac Newton, who was born in 1642, gravity and orbital motion were thought to be unrelated. Newton proposed, however, that the Moon would fly off in a straight line if there were no force keeping it from doing so. He recognized that there was a force of gravity constantly pulling the Moon toward the Earth. Taking this discovery one step further, he understood that gravity was a universal phenomenon— between any two bodies there is an attracting force. He also recognized that gravity increases with more massive bodies and diminishes as bodies move farther apart.

Newton determined the physical laws that explain Kepler's observations and laws of planetary motion. He demonstrated that the total gravitational attraction of a spherical body is exactly the same as if all its mass were concentrated at the center of the sphere. Building on Kepler's discovery that orbits could be elliptical, Newton showed that they could also be circular, parabolic, or hyperbolic. This provided an explanation for the behavior of certain comets that do not follow elliptical paths. Newton also proposed that since any object that has mass is attracted to every other object that has mass, the orbit of each planet is influenced not only by the Sun, but by all the other planets as well. In general the influence of the other planets is very small because of the far greater mass of the Sun.

Late in his life, Isaac Newton said he had been "like a boy playing on the seashore, and diverting myself in now and then finding a smoother pebble or a prettier shell than ordinary, whilst the great ocean of truth lay all undiscovered before me." His "pretty shells" were ideas that changed the course of modern science.

Discovering New Planets

Until the last part of the eighteenth century, only six planets of the solar system were known—Mercury, Venus, Earth, Mars, Jupiter, and Saturn. Using steadily improving instruments, astronomers made increasingly precise observations, checking each finding against Newton's laws. Then, in 1781, William Herschel, a British astronomer, discovered Uranus. In 1846, Urbain Leverrier in France and John Couch Adams in England independently concluded that the *perturbations,* or disturbances of motion, in the orbit of Uranus could not be explained by the presence of other known bodies, and that an eighth planet must exist. Asked by Leverrier to watch a certain area of the sky for an uncharted object, the Berlin Observatory found the predicted planet, which was named Neptune.

For the same reasons, another planet beyond Neptune was thought to exist. In 1905 the search began when Percival Lowell and William Pickering predicted its orbital characteristics and location. A wide-field camera was used, and in 1930 Clyde W. Tombaugh found the new planet within 6° of where Lowell predicted it would be located. It was named Pluto. Although the existence of the nine planets of the solar system was now known, the nature of these worlds remained a mystery to be solved.

Donato Creti, an Italian painter at the turn of the eighteenth century, captured the sense of human wonder and curiosity about the solar system in his work "Astronomical Observations."

Modern Views of the Solar System

With the planets in our solar system identified, the search for answers to larger questions continues: What forces caused the birth of the universe, the formations of the galaxies and stars, and our solar system? Where does our solar system fit in the greater cosmic picture? Investigations continue, but some generally accepted scientific conclusions have been reached.

The Origin of the Universe

To date, the most widely accepted theory of the origin of the entire universe is still the Big Bang theory. It proposes that all the matter in the universe came from a specific event about 10 billion years ago, when an incomprehensibly dense and compact mass exploded. This spread the first matter—hydrogen and helium—throughout the universe.

As the temperature dropped, clouds of matter condensed and collected due to gravitational attraction, and galaxies and stars formed. A star is born when matter accumulates and the pressure and temperature in the core increase. When the mass reaches about 8 percent of the mass of our Sun, a nuclear reaction takes place involving the fusion of hydrogen atoms to produce helium. The sustained nuclear reaction produces light and heat for a prolonged period of time.

Where Are We?

It is difficult for us to comprehend just how small our Milky Way galaxy is in relation to the entire universe and how small our solar system is within that galaxy. Our galaxy is one of countless billions of galaxies, and our Sun is only one of about 300 billion stars within the Milky Way galaxy. The solar system is located roughly halfway between the center and the edge of the galaxy, moving around the galactic center in a nearly circular orbit once every 240 million years.

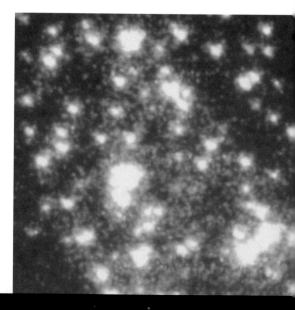

A very massive star may rapidly evolve to a stage where it explodes in a supernova. A supernova creates elements heavier than iron and releases elements such as oxygen, silicon, carbon, nitrogen, aluminum, and iron from the star's interior. Supernova explosions spread this matter throughout the galaxy, together with the hydrogen and helium already present. Today, about 76 percent of the matter in the universe is hydrogen, and most of the rest is helium, with only small amounts of the heavier elements.

Formation of Our Solar System

There is currently no single theory for the origin of the solar system that can account for the diverse nature of its bodies. It is generally accepted, however, that our solar system began to form about 4.5 billion years ago from a vast hot cloud of gas and dust. Perhaps triggered by a supernova, the cloud, or *nebula*, began to collapse and rotate around a center of mass.

When the central mass reached about twice the mass of present-day Jupiter, the pressure in the interior made it so hot and dense that it became a *protostar*, or precursor to a star—our Sun. At the same time, adhesive forces, such as electrostatic charge, began to form grains from the nebular dust. Grains grew into clumps and the clumps grew into larger bodies perhaps as large as our Moon.

Nuclear fusion began when the protostar reached a mass about 80 times that of present-day Jupiter. The shock wave blew away remaining uncondensed nebular gases. Gravitational interactions caused the bodies to collide and grow by accretion, or accumulation of matter. These larger bodies eventually became the planets.

The composition of each planet was controlled by the composition of the dust where the planet formed. Because the temperature decreased with distance from the Sun, the composition of the inner planets included silicates and metals that condense at high temperatures. The composition of the colder outer planets included ices made of methane, ammonia, and water.

Set against a background image of stars, The Milky Way galaxy (left) includes our solar system, at a distance of about 25,000 light-years away from its center.

This picture of asteroid 243 Ida was made from images taken by the *Galileo* spacecraft in August 1993. *Galileo* used infrared filters, making the asteroid look more reddish than it would appear to the unaided human eye. The sequence of images used to create the picture reveal that Ida has its own moon, which appears to the right of the asteroid.

Asteroids and Meteorites

Much of the leftover debris from the formation of the solar system is in orbit around the Sun in the region between Mars and Jupiter known as the *asteroid belt*. There the gravitational pull of the planet Jupiter was too great to allow another planet to form by accretion. Asteroids can be found anywhere, from inside Earth's orbit to beyond the orbit of Saturn. Many are so large they resemble minor planets.

A group of as many as 2,000 asteroids have very elliptical orbits that cross Earth's orbit. These Earth-crossing asteroids like those in the main-belt are heavily cratered, reflecting long collisional histories. Fragments of stone or iron ejected from the asteroids by impacts (also bits of debris from comets) are called *meteoroids*. A meteoroid entering Earth's atmosphere forms a streak of light called a *meteor* or shooting star. Fragments large enough to survive the fall through the atmosphere are called *meteorites*.

Some meteorites are samples of the most ancient material in the solar system, unchanged from the time the planets began to accrete. Thus, they reveal much of what we know about the composition and conditions in the solar nebula when the planets were forming, as well as the age of the solar system.

A continuous rain of meteoritic material falls to Earth. Most meteorites are small pebble-size objects that barely survive burning up in the atmosphere and often land unnoticed. Stone-size meteorites are not uncommon but are more rare. Larger meteorites whose weights are measured in tons are the exception. One iron meteorite weighing 60 tons (54 MT) fell in Africa. Another weighing 34 tons (31 MT) landed in Greenland. In February 1969 a stony meteorite broke up, producing a spectacular meteorite shower that occurred in Mexico, with collected specimens totaling about 2 tons (1.8 MT).

Meteorites from the Moon and Mars

This 15–pound (6.7-kg) meteorite, found in Antarctica, is estimated to be 1.3 billion years old. Its composition indicates that it may be a fragment of Mars.

Meteorites are rocks that fall through Earth's atmosphere and strike the ground. Scientists can determine the age of these rocks by measuring the rate of radioactive decay of certain elements in them. This reveals how much time has passed since the rock was "born" in a process of melting, crystallizing, and cooling. Most meteorites have been found to be about 4.5 billion years old—as old as the solar system itself.

However, a few meteorites have been discovered that are surprisingly younger—only about 1.3 billion years old. Because asteroids are small and geologically inactive, no relatively young rocks should have formed on these bodies. Thus, the young meteorites must be fragments of larger planetary bodies, which have their own internal sources of heat.

In 1979 a controversial theory was suggested: these younger meteorites were debris from a large impact on Mars. Most scientists believed, however, that any impact forceful enough to accelerate a fragment to its escape velocity—the speed at which it could escape the planet's gravity—would pulverize or melt the fragment. Even the escape velocity of the Moon was thought to be too great.

In 1982 the chemical makeup of a meteorite found in Antarctica matched the Moon rocks returned by the Apollo crew. In the following years, several more lunar meteorites were identified, reviving the theory that the young meteorites may also have come from Mars. Since then, one of the young meteorites from Antarctica was found to contain inert gases in the same proportions as the *Viking* spacecraft had measured in the Martian atmosphere. By the end of the 1980s many scientists agreed that samples of the lunar and Martian crust had landed on Earth as meteorites.

The Planets

The nine planets orbit the Sun on almost the same plane. Seen from Earth, they appear to move across the sky through a rather narrow zone, the so-called zodiac. Observed from a position far above Earth's north pole, all the planets appear to orbit the Sun in a counterclockwise direction. All the planets except Venus appear to rotate or spin in the same counterclockwise direction.

Because of their histories and locations, the inner planets—Mercury, Venus, Earth, and Mars—differ from the outer gas giants, often called the "Jovian planets," after Jupiter. The inner planets are smaller, denser, and rockier. What atmospheres they have are due to internal activity and are not related to the masses of hydrogen and helium that originally enveloped them. They are also warmer and rotate more slowly than the outer Jovian planets.

Jupiter, Saturn, Uranus, and Neptune are much less dense than the inner planets. The bulk of their masses consists of hydrogen and helium, with some methane and ammonia. These rapidly rotating planets are cold and icy with deep atmospheres and ice-rich moons.

Pluto seems to be a misfit. The farthest out, it is small and ice-rich, with little atmosphere, and spins slowly. It probably has a unique history that accounts for its differences from the other outer planets.

This *orrery*, a mechanism showing the relative positions and motions of the planets and their moons, was built around 1828 by A. Willard Jr. of Boston. Since then, Uranus, Neptune, and Pluto have been discovered by astronomers.

Modeling the Solar System

It would be helpful to see a single scale model of the solar system that would show the relative sizes of the planets, the divergence of their paths from the plane of the Earth's orbit, the degree to which each is tipped out of vertical, the shape of each planet's orbit, and the distances between those orbits. Unfortunately, the vast differences in sizes and distances among the planets make this very difficult.

Some idea of the relative sizes and distances of the planets can be gained by imagining that the Sun is located in New York City and Pluto in San Francisco. Using the distance between these two cities as a scale, the Sun would be over one-half mile (980 m) wide. It would just fit inside Central Park and be more than three times taller than the World Trade Center. Mercury, measuring 11 feet (3.4 m) in diameter, and Venus, 28 feet (8.5 m), would be located in New Jersey. Earth would have a diameter of just under 30 feet (9 m) and be found on the New Jersey-Pennsylvania border. Mars would be in eastern Pensylvania, near Wilkes-Barre. At 328 feet (100 m) across, Jupiter would just fit between the goals of a football field in Erie, Pennsylvania. Saturn would be in Michigan, Uranus in Nebraska, and Neptune in Utah. Pluto, at a little over 5 feet (1.6 m) in diameter, might go unnoticed in San Francisco.

The Sun

The Sun was worshiped as a god
by early cultures. Structures were built
in its honor and rituals performed to
secure its favors. Such practices and
beliefs faded over the centuries as
astronomers gained knowledge about
the sphere whose nuclear furnace
cradled and continues to support life on
Earth. Not until the twentieth century,
however, were the Sun's complex nature
and many of its fascinating features
revealed.

The Sun, source of
the energy that makes
life on Earth possible,
appears as a glowing
sphere suspended over
the ocean.

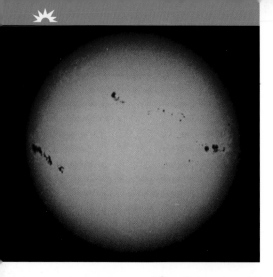

If an airplane could fly to the Sun at the speed of sound, which is about 740 miles per hour (1,190 km/hr), it would take 14 years to get there.

In this image the surface of the Sun that is visible to the human eye is seen through a special filter that transmits a narrow band of light near the color orange. Several sunspots are visible, and the center of the Sun appears brighter where the temperature is higher.

The Sun as a Star

The Milky Way galaxy, home to our solar system, contains about 300 billion stars. Of all the stars in our galaxy, only our Sun is close enough for us to study it in detail. The Sun is located about halfway from the center of the Milky Way galaxy, toward its edge.

Most known stars have masses between one-tenth and ten times the mass of the Sun, so the Sun can be considered average size. It is also of average age for a star—about 4.5 billion years old—and about halfway through its life cycle consisting of about 10 billion years.

Measuring Distances in Space

To help describe the vast distances of the solar system, astronomers use the *astronomical unit,* abbreviated *AU.* One AU is equal to 93 million miles (150 million km), the average distance between the Sun and Earth. Jupiter, for example, is 5.2 AU from the Sun, which means that it is 5.2 times farther from the Sun than Earth.

It takes light 8 minutes to travel a distance of 1 AU. Thus sunlight reaching Earth at any moment left the Sun's surface 8 minutes earlier.

Light-Year Even the astronomical unit is too small for dealing with distances outside our solar system. The unit commonly used to describe such vast distances is the *light-year.* Light travels at the rate of 186,000 miles per second (300,000 km/sec), so a light-year, or the distance light travels in a year, is about 6 million million miles (9.5 million million km). The Milky Way galaxy is about 100,000 light-years across. The next nearest star in our galaxy, Proxima Centauri, is about four light-years away. In a scale model with the Sun and Earth one foot apart, Proxima Centauri would be 51 miles (82 km) away.

The Ulysses Sun Satellite

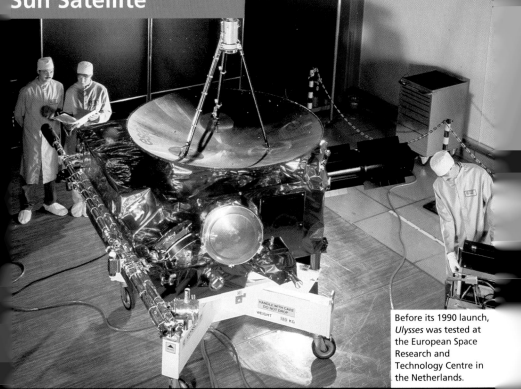

Before its 1990 launch, *Ulysses* was tested at the European Space Research and Technology Centre in the Netherlands.

A radiation-resistant spacecraft weighing about 810 pounds (370 kg), *Ulysses* was launched in October 1990 onboard the Space Shuttle *Discovery*. As it flew by Jupiter in February 1992, the gravitational field of that planet placed *Ulysses* on its course to polar orbit around the Sun where it was to complete its primary mission: accurate assessment of the total solar environment. It is the first probe to explore the environment of the Sun over its poles.

On November 5, 1994, the Jet Propulsion Laboratory (JPL) team tracking the mission for the National Aeronautics and Space Administration (NASA) reported that *Ulysses* had completed the initial phase of its mission—a pass over the southern solar pole, about 70 degrees south of the Sun's equator. The spacecraft then proceeded back toward the Sun's equator and toward its north pole.

Initial measurements returned from *Ulysses* indicate that the Sun has a uniform magnetic field rather than solar magnetic poles. The probe also revealed that in the southern polar region, the solar wind is blowing at about 2 million miles per hour (3.2 million km/hr). This is nearly double the speed at which the wind flows at the Sun's lower latitudes.

Ulysses is now trapped in its polar orbit around the Sun. Its exact south-to-north polar passes in 1994 and 1995 will be repeated in 2000 and 2001, respectively. This second set of polar passes will take place when the sunspot cycle will be at its maximum, allowing *Ulysses* to measure a wide range of solar activity.

Prominence

NASA's *Skylab 4* took this background photograph of the Sun in December 1973, showing a spectacular solar flare (upper left) spanning more than 376,000 miles (588,000 km). The cross section of the Sun shows the approximate proportions of its interior.

The Composition of the Sun

A huge ball of gas, the Sun contains 99.9 percent of the mass of our entire solar system. About 330,000 times as massive as Earth, it could contain all the planets a hundred times over. Its diameter of 863,700 miles (1,390,000 km) far exceeds Earth's diameter of 7,926 miles (12,756 km). It is composed of about 92 percent hydrogen, 7.8 percent helium, and less than 1 percent heavier elements, such as oxygen, carbon, nitrogen, and neon.

Like the planets, the Sun rotates, but due to its gaseous composition, different parts rotate at different rates. This is known as *differential rotation*. The equatorial regions rotate fastest, in a period of 25 days, and the poles the slowest, in 36 days.

Solar Energy

All of the planets produce energy either from the decay of radioactive elements or from gravitationally induced contraction. But the Sun produces a hundred million times more energy than all the planets combined. Just over half the Sun's energy takes the form of visible light—that part of the electromagnetic spectrum that the human eye can see. Most of the rest takes the form of infrared radiation. The tiny fraction of the Sun's energy that reaches Earth amounts to only about a billionth of its total energy output.

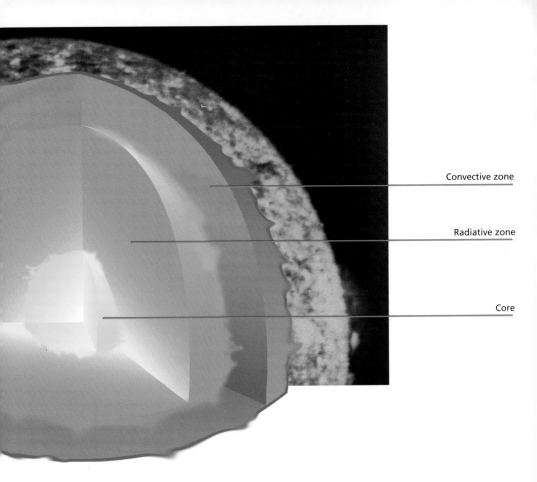

Convective zone

Radiative zone

Core

The Solar Core

The principal difference between our Sun and other stars and the outer planets, known as gas giants, is their mass. Stars have a much greater mass, and thus the pressure and temperature in their interior is much greater because of the weight of their overlying mass. If, as in the Sun, the material in the core is primarily hydrogen, this huge weight causes atoms of hydrogen to combine or fuse into helium, releasing energy. This nuclear fusion occurs on an enormous scale, producing vastly larger amounts of energy than any planet produces. A star is prevented from collapsing on itself because the inward gravitational pressure due to its mass is exactly balanced by the outward pressure of expanding gases heated in the core by nuclear fusion. Because the pressure exerted by the gas is directly related to the temperature of the star's core, the temperature in the Sun's core is estimated to be about 27,000,000°F (15,000,000°C).

How We Know: Spectrographs

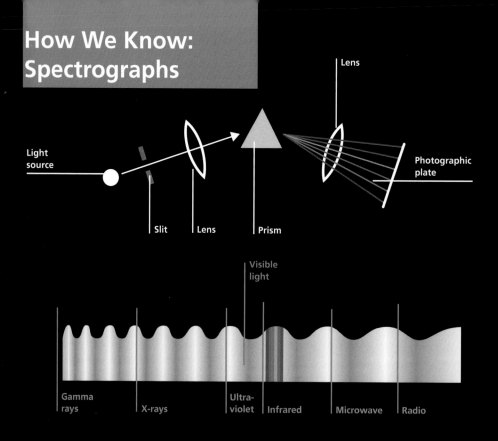

Light source

Slit

Lens

Prism

Lens

Photographic plate

Visible light

Gamma rays

X-rays

Ultra-violet

Infrared

Microwave

Radio

Using an instrument called a spectrograph, astronomers have been able to determine what chemical elements make up the atmospheres of the planets. These experiments were carried out long before any probe was sent to analyze the planets' gases.

A spectrograph makes use of Isaac Newton's early research into the properties of light. Using a prism, Newton discovered in 1665 that sunlight is not really white but comprises many different colors of light. A spectrograph separates light into its component colors or wavelengths and creates a corresponding map of the visible portion of the electromagnetic spectrum.

Every chemical element gives off specific wavelengths of light when it is heated. Thus, each element has a unique spectrum, a "fingerprint" that can be used to identify it. When a gas receives light from a source that is hotter than itself, the gas absorbs the exact same wavelengths that it would emit if it were heated. The wavelengths that are absorbed show up as dark lines on the spectrograph. Analyzing the location of these dark lines in the spectrum reveals the chemical elements present in a planet's atmosphere. A spectrograph has also been used to determine the chemical elements of the Sun's atmosphere and those of distant stars.

The core, where fusion takes place, occupies only 1.5 percent of the volume of the Sun. Because of the tremendous pressure, the core is eight times denser than gold, even though it is composed of gases.

The Fusion Process In the fusion process four hydrogen nuclei combine to make one helium nucleus. The mass of the new helium nucleus is 0.7 percent less than the mass of the four hydrogen nuclei that combined to make it. Every second, the Sun processes 700 million tons (635 million MT) of hydrogen into 695 million tons (630 million MT) of helium. Most of the mass lost is converted to energy. This means that 5 million tons (4.5 million MT) of matter is converted into energy every second.

Luminosity The luminosity of a star is the measure of the total amount of radiation it produces every second. It is also an indication of its rate of energy production. This rate in turn depends on the core pressure and temperature of the star, which depend on its mass. During its lifetime the Sun has used up about half its hydrogen atoms. This reduction in the number of atoms lowers the outward pressure. In order to keep the inward pressure from the weight of the overlying gases and the outward pressure in balance, the core temperature and rate of fusion must increase. Thus, over its 4.5-billion-year life, the luminosity of the Sun has increased by about 40 percent, and it will continue to increase for some time.

Energy Transmission

Energy released in the Sun's core moves slowly outward toward the surface until it finally emerges several hundred thousand years later in the form of visible light and infrared radiation. Energy passes from the core to the surface in two ways—by *radiation* and by *convection*.

Energy generated in the core is carried outward by radiation. The extreme heat in the interior makes it more *transparent* to radiation, meaning more energy can pass through it without being absorbed. This is true for about 70 percent of the distance from the core to the surface of the Sun.

By this point, the temperature has dropped 90 percent, from 27,000,000°F (15,000,000°C) to 2,700,000°F (1,500,000°C). At this lower temperature, the gases are more opaque to radiation, thus more energy is absorbed. The energy travels the remaining 30 percent of the way by convection. Convection involves overturning of hot gases. These gases rise to the surface in eddies, cool, then sink back into the interior.

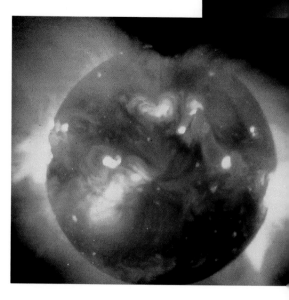

The tremendous turbulence of the Sun's upper atmosphere is revealed in this image, taken in 1992 by an x-ray telescope on the *Yohkoh* spacecraft, a joint Japanese, American, and British mission. *Yohkoh* means "sunbeam" in Japanese.

How to View the Sun Safely

The Sun is such a concentrated source of light that looking at it directly can cause permanent eye damage. The temptation is particularly hard to resist during a solar eclipse. There are, however, ways to view the Sun without risking your eyesight. One way is to use a telescope to project an image of the Sun onto a sheet of white paper. Or—even simpler—use a pinhole in a piece of cardboard to project the image.

The Atmosphere of the Sun

The Sun's atmosphere is divided into two layers: the *chromosphere*, and the *corona*. Below these layers lies the *photosphere*.

The Photosphere

The visible surface of the Sun is called the photosphere ("light sphere"). It is a thin shell of gases about 125 miles (200 km) deep. The vast majority of the energy radiated by the Sun comes from the photosphere after making its way to this shell from the interior. It has a temperature of about 10,000°F (5,500°C). The density of the photosphere decreases abruptly at its outer limit, giving it the appearance of a sharp edge as seen from Earth.

Although the surface appears smooth to us, it is actually turbulent, reflecting vigorous convective motion. Recently, features known as *granules* were discovered. These are the tops of convective elements that bring the energy to the surface. An individual granule is roughly 620 miles (1,000 km) across, or about the size of Texas. A granule can rise upward as fast as 1,100 miles per hour (1,800 km/hr). Another, more familiar feature of the photosphere is the *sunspot*.

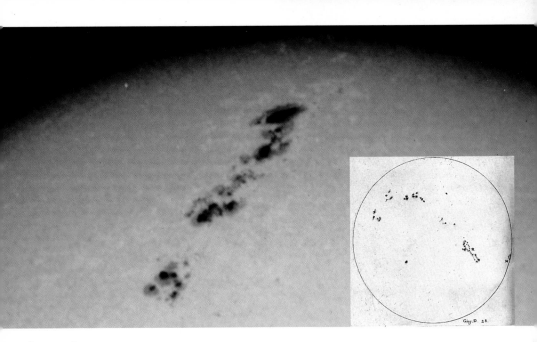

Sunspots

Dark spots have been observed in the photosphere of the Sun for thousands of years. Some sunspots are as large as Earth, and individual sunspots may last for several weeks. In visible light, sunspots appear dark against an otherwise bright photosphere. They appear dark because they are cooler than the rest of the photosphere—about 6,300°F (3,500°C) as compared to 10,000°F (5,500°C). They radiate only 20 percent as much energy as equal areas of the rest of the photosphere.

The Sun has very strong magnetic fields, thought to be generated by convective motion and differences in the rates of rotation of the Sun's gases. These differences (about 25 days at the equator and about 36 days at its poles) appear as deep as at least 124,000 miles (200,000 km). Deeper than that, the rotation period seems to be a stable 27 days.

Sunspots are also known to have strong magnetic fields associated with them. In fact, the Sun's magnetic field is so strong in the area where a sunspot occurs that convective motion is greatly reduced beneath it. Since this in turn reduces the amount of heat brought to the surface as compared to the surrounding area, the spot becomes cooler.

The two distinct regions of sunspots are visible in this image of the solar photosphere: the dark central region is the *umbra*, and the surrounding region is the *penumbra*.

Galileo used his telescope to project sunspot patterns onto a paper screen. Their shifting motion proved to him that the Sun rotates on its axis.

The position and number of sunspots varies during an 11-year cycle. At the beginning of a new cycle there are few sunspots and they form at mid-latitudes. As the cycle progresses the number of sunspots increases and they form at lower latitudes.

While the 11-year cycle has been consistent since 1710, almost no sunspots were recorded from 1640 to 1710. This period, referred to as the Maunder Minimum, has never been explained. It did coincide with the "little ice age" in northern Europe, a period of below-average temperatures, but no direct connection has been established.

The Chromosphere

The chromosphere ("color sphere") is a thin layer of gases only a few thousand miles thick that lies just above the photosphere. During a solar eclipse, just after the Moon covers the Sun, the chromosphere appears as a red outline. It is much hotter than the photosphere. From a temperature of 7,600°F (4,200°C) at its inner edge, it reaches 14,800°F (8,200°C) near its outer edge. Higher up, there is an abrupt transition zone where the gases can no longer cool themselves efficiently. This is in the realm of the corona, and the temperature there zooms to more than 1,800,000°F (1,000,000°C).

Data supplied by NASA's *Solar Maximum Mission* satellite provided this colorful view of the solar corona. Colors representing the densities of the solar corona range from purple (densest) to yellow (least dense).

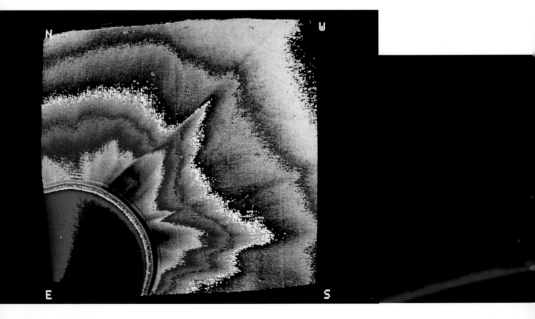

The Corona

The top layer of the Sun's atmosphere is called the corona ("crown"). It appears as a pale white glowing area around the Sun during a solar eclipse. The corona's glow is caused by sunlight scattered by electrons freed from atoms by the high temperatures. The high temperature in the corona is thought to be caused by the interaction of gas with the photosphere's strong magnetic fields. The corona extends millions of miles into interplanetary space.

Prominences and Flares

Observed on the edge of the Sun, a *prominence* is a mass of ionized gas carried from the surface into the corona, often forming a loop or series of loops. An ionized gas conducts electrical charges because its atoms have been stripped of their electrons. These atoms respond to a magnetic field like iron filings following a magnet's lines of force. The ionized gases are held by magnetic lines of force connected to sunspots. Occasionally one end of a prominence breaks loose from the Sun, catapulting its gas away from the Sun and into space. It is then called an *eruptive prominence*.

A solar *flare* is perhaps the most spectacular feature associated with the activity of the Sun. It releases enormous amounts of energy into space, sometimes accompanied by large amounts of gas.

Solar flares that are accompanied by a surge of rising gas are shown in this image. In 1989 a series of violent solar flares knocked out electricity all across the Canadian province of Quebec.

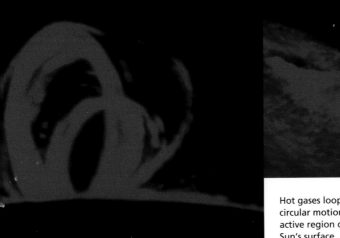

Hot gases loop in a circular motion above an active region of the Sun's surface.

Solar Eclipses

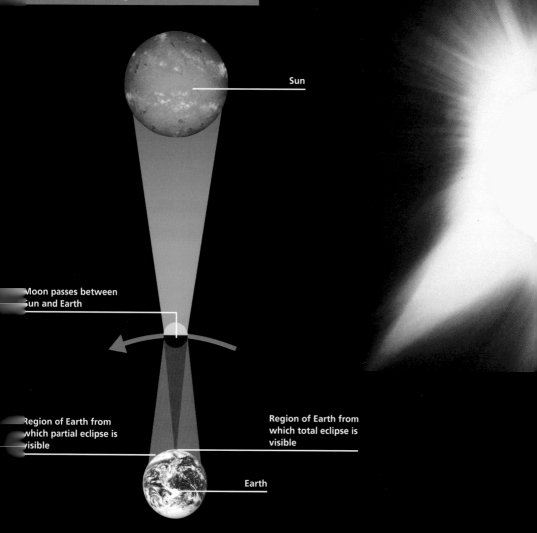

Sun

Moon passes between
Sun and Earth

Region of Earth from
which partial eclipse is
visible

Region of Earth from
which total eclipse is
visible

Earth

Solar eclipses have re-sulted in some interesting historical events. For example, a total solar eclipse occurred during a battle between the Lydians and Medes of Asia Minor in 585 B.C. The fright-ened soldiers then ceased fighting and signed a peace treaty.

Ancient Egyptians believed an eclipse occurred when an underworld serpent swallowed the boat carrying the sun-god Ra, during his daily journey across the sky. Other cultures associated solar eclipses with outbreaks of disease. Written records dating back to 763 B.C. from Babylonia and China, however, indicate that many ancient astronomers understood the cause of an eclipse.

A solar eclipse takes place every 18 months when the Moon passes between Earth and the Sun. The Sun projects the Moon's shadow onto Earth. People who live where the shadow falls see a total solar eclipse. Because both the Sun and Moon are moving, the shadow traces a path

During a total eclipse, the Sun's corona, which is not normally visible, can be seen. Here coronal streamers surround the Sun.

Just before and just after totality, beads of sunlight create the "diamond ring effect" around the Moon.

across the Earth's surface. People near this path see a partial solar eclipse.

During a total eclipse the Moon slowly moves in front of the Sun and the sky darkens. Stars and planets appear and animals may sleep, mistaking the dark sky for night. When the Moon nearly covers the Sun, bright beads of light, known as Bailey's beads, are visible. These are places where the Sun shines through the Moon's craters and valleys.

The Sun is 400 times larger than the Moon and the Moon is 400 times closer to Earth, so the Sun and Moon appear to be the same size. This allows the Moon at times to completely cover the Sun during an eclipse. Totality can last from a few seconds to several minutes.

Total solar eclipses seem rare because 300 years can pass before one is seen in the same location again. When the Moon is farther from Earth in its orbit, it does not completely cover the Sun. The result is a partial or annular eclipse.

Flares can release energy equivalent to more than a billion one-megaton thermonuclear explosions in a few seconds and may eject as much as 10 billion tons (9 billion MT) of matter. Solar flares are thought to result from violent explosions in the solar atmosphere that are triggered by the sudden release of energy from tangled and twisted magnetic field lines, much like a rubber band that has been stretched and twisted, then released. The energy particles reaching Earth from solar flares interact with Earth's magnetic field and cause the aurora borealis. They also may interrupt radio transmission and induce currents in electrical grids large enough to overload transformers, and they can even cause confusion in the navigation of birds.

Solar Wind

Auroras are touched off when charged particles resulting from solar flares interact with Earth's magnetic field and atmosphere. Here colorful, shimmering rays of the aurora australis, or southern lights, flicker and pulsate across the sky. This image was taken during a Space Shuttle *Discovery* mission.

A constant flow of highly ionized gas called *solar wind* emanates from the Sun and sweeps the solar system. When the magnetic field forms arches that extend high in the corona, these lines weaken as they move away from the Sun, forming holes in the corona. Here the pressure of the gases is strong enough to allow them to escape. The gases form the solar wind, which rushes up through the holes in the corona and spreads throughout the solar system. About five days after gases have escaped, the high-speed stream of the solar wind reaches Earth and can be detected by orbiting spacecraft. The solar wind may reach a velocity of more than 435 miles per second (700 km/sec) and have a density of from 10 to 100 particles per cubic centimeter.

Solar Sailing

A spacecraft on a long interplanetary mission could possibly be powered by a solar sail. Once in orbit, the spacecraft would unfurl a lightweight, aluminized plastic sail. Pressure on the sail from sunlight would push the craft, while changing the position of the sail would alter the speed. The Sun would be an economical power source on journeys through the planetary system that might last for years.

The Future

The Sun and our solar system, like all other things, are destined to change, but not anytime soon. Solar scientists have made many predictions about the future of the Sun. Throughout the next 5 billion years, the Sun's radius, luminosity, and temperature will continue to increase. In about 6.5 billion years, the Sun will have expanded to more than 3.3 times its present size. It will look like a swelling, orange-red disk, because most of the hydrogen in the core will be gone, forcing it to consume hydrogen farther out. By that time, the temperature on Earth will be a great deal hotter than it is now, and all water on the planet will have vaporized.

The Sun will then expand to 100 times its present size and become what astronomers call a *red giant*, engulfing Mercury. When the Sun's core reaches a temperature of about 180 million °F (100 million °C), helium fusion will be triggered, generating carbon and oxygen. The explosion caused by this new level of fusion will blow away as much as a third of the Sun's body.

Eventually the Sun will lose its outer layers and become what is known as a *white dwarf*, about the size of Earth. Essentially, this white dwarf will be the tremendously massive, and dense, original core of the Sun. The remains of the planets and their satellites will continue to orbit in near darkness.

The Sun, Earth's parent star, is estimated to be about halfway through its life span of approximately 10 billion years. As it rises and sets, the Sun provides some of the most beautiful spectacles on our planet.

The Inner Planets

Mercury, Venus, Earth, and Mars—the four planets closest to the Sun—share a common history in the solar system. Unlike the outer gaseous planets, these terrestrial planets were formed from dense, rocky materials. Yet despite many basic similarities, they are sharply different worlds. From the intensely cratered surface of Mercury to the highlands of Venus to the giant volcanoes of Mars, each offers clues to the story of our own Earth and its place in the scheme of planetary evolution.

Data gathered by the *Magellan* spacecraft were used to produce a computer-generated view of the 5-mile-high Maat Mons volcano on Venus.

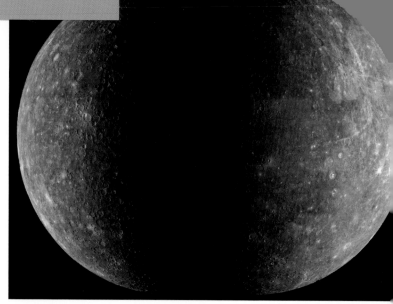

These photo-mosaics of Mercury were compiled from images returned by the *Mariner 10* spacecraft as it approached and moved away from the planet. These images show the varying range in size of Mercury's thousands of craters.

Mercury

Named for the swift messenger god of Roman mythology because of its speedy orbit around the Sun, Mercury is the fourth brightest planet to the naked eye (after Venus, Mars, and Jupiter) and the second closest to Earth. However, the fact that it is the closest planet to the Sun makes observation of Mercury difficult from Earth.

The best times to view Mercury occur only twice a year when it appears above the horizon at its greatest angular distance from the Sun. At these times, Mercury can be seen just after sunset or just before sunrise, but even then it must be viewed through the greatest thickness of Earth's atmosphere. For these reasons, Mercury was a difficult planet to study until radar observations and robotic space probes allowed a closer look.

Size, Gravity, and Density

About one-third the size of Earth and slightly larger than the Moon, Mercury is the second smallest planet, next to Pluto. Its diameter is 3,031 miles (4,878 km). It has about the same gravity as Mars and a little more than one-third the gravity of Earth. But Mercury's average density of 5.4 (the measurement of the amount of mass in a given volume of space; see page 54) is second among the planets only to Earth's average density of 5.5. It is thought that a huge impact may have removed part of Mercury's outer layer, leaving a higher proportion of the denser core than other planets have. By weight, Mercury seems to be as much as 70 percent iron and only about 30 percent rock.

Mercury's History

Mercury formed at the same time as the other planets, about 4.5 billion years ago. Because Mercury formed close to the Sun, it must have been subjected to very high temperatures, possibly high enough to completely melt the planet. Elements with high densities, particularly iron, completely separated from the molten rock and sank toward the planet's center. This resulted in a large iron core and a mostly iron-free mantle and crust, composed mostly of silicates (compounds of silicon and oxygen). Mercury's core is believed to comprise about 75 percent of its total diameter, a much larger percentage than in the other terrestrial planets.

Discovery Rupes (above), the largest known scarp on Mercury, appears as viewed by *Mariner 10*. This scarp reflects crustal shortening due to global contraction of the planet.

Cratering, Volcanism, and Tectonics

Three primary processes have shaped many of the surface characteristics of Mercury and the other terrestrial planets, including Earth. *Impact cratering* occurs when colliding bodies create bowl-shaped cavities. *Volcanism* occurs when molten, or melted, material from the interior of a planet or moon reaches the surface. *Tectonics* is the movement of a planet's crust. The results of all three of these processes are evident on Mercury.

For the first 600 million years of their existence, the four inner planets were continuously bombarded, leaving craters of widely varying sizes. During this period of heavy bombardment, the interior of Mercury was melting and the planet was expanding. The melting led to flood volcanism, in which enormous amounts of molten lava poured out at the surface, covering large areas and forming plains between craters. Flood volcanism is common to

Mariner 10 made three flybys of Mercury in 1974 and 1975, providing data about the structure of the planet. The Caloris Basin (right) is the largest known impact basin on Mercury—comparable to the largest impact basins on the Moon or Mars.

all four inner planets but differs from the type of volcanism that produces steep-sided stratovolcanoes like Mount Rainier in Washington State or broad, rounded, shield volcanoes like Mauna Loa on the island of Hawaii.

Sometime near the end of the bombardment, about 4 billion years ago, an asteroid estimated to be about 62 miles (100 km) in diameter struck Mercury. When an object this large hits a planet, it creates an impact basin—in this case, the Caloris Basin, one of the largest in the solar system. Measuring about 810 miles (1,300 km) across, the Caloris Basin could hold the entire state of Texas. As is typical with large impacts, lava flooded the basin. The basin floor contracted and stretched as it settled under the weight of the volcanic material, forming fractures and ridges. The impact generated seismic waves that may have gone through and around the planet, breaking the surface on the opposite side into blocks and depressions to form Mercury's hilly terrain.

As Mercury continued to cool, it began to contract. At the same time small objects continued to strike the surface, giving the planet a cratered surface like the Moon's. In fact, the distributions of crater sizes on Mercury, the Moon, and Mars are very much alike, suggesting that the same family of objects was responsible for the heavy bombardment experienced by all the terrestrial planets during this period.

When Mercury cooled and contracted, its crust adjusted, resulting in global compression and tectonic activity. This tectonic activity created numerous *scarps,* or cliffs. The largest scarp observed to date is Discovery Rupes, about 310 miles (500 km)

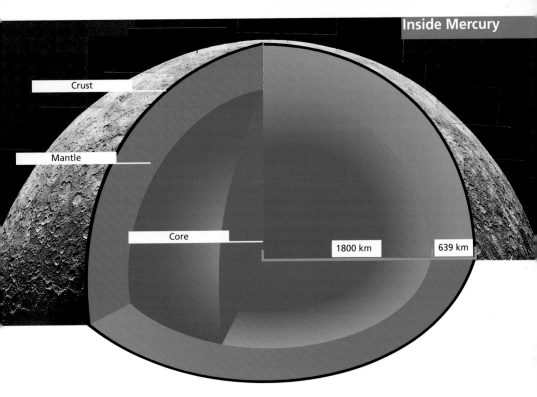

Crust

Mantle

Core

1800 km

639 km

long and almost 2 miles (3 km) high. The shortening of the crust meant that Mercury's circumference decreased by as much as 1.25 miles (2 km).

Many of Mercury's large craters are not covered by lava flows, indicating that by the time the heavy bombardment of the inner planets ended—about 3.8 billion years ago—most volcanic activity on Mercury had ceased as well.

A cross section of Mercury shows the relationship of its major parts. Many kinds of information are used to estimate a planet's interior structure. In the absence of seismic data, the mean planetary density is an important clue.

Mercury's Atmosphere

Mercury has almost no atmosphere. The thin atmosphere that does exist consists primarily of helium and sodium, much of which may be the product of solar wind. The atmospheric pressure is only a million-billionth that of Earth's—as low as many vacuums created in Earth laboratories.

Exploration of Mercury

Our first close-up views of Mercury were provided by the *Mariner 10* spacecraft, which was equipped with two television cameras that returned digital images. *Mariner 10* flew by Mercury on March 29, 1974, then orbited the Sun and made two more fly-bys—one on September 21, 1974, and one on March 16, 1975. In those three passes it photographed about 45 percent of the planet's surface, revealing a landscape similar to our Moon's.

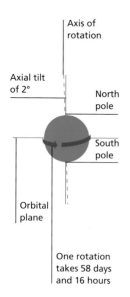

Axis of rotation

Axial tilt of 2°

North pole

South pole

Orbital plane

One rotation takes 58 days and 16 hours

Mercury's Orbit and Rotation

Mercury has the shortest year of any planet. While it takes Earth approximately 365 days to orbit the Sun, it takes Mercury only about 88 Earth days to complete one orbit around the Sun.

Mercury's orbit is the most elliptical of any planet other than Pluto. At its closest orbital position to the Sun, called *perihelion,* Mercury is 29 million miles (47 million km) away. At its farthest orbital position from the Sun, called *aphelion,* it is 44 million miles (71 million km) away. At perihelion it receives more than twice the heat that it receives at aphelion.

Before 1962 Mercury was thought to rotate only once during each of its orbits, like the Moon. If this were true, the same side would always be turned toward the Sun. Later radar observations showed that the rotation period is about 59 Earth days. Thus Mercury rotates exactly three times while it orbits the Sun twice, so in two of its years the planet has completed only three of its days.

Strange Sights Mercury's rotation and highly elliptical orbit produce some sights that would seem quite peculiar to a visitor from Earth. From some locations on Mercury during a single day, the Sun would appear to halt its east-west progress, turn back toward the east for a short distance, then reverse direction again to resume its westerly journey. At other locations, the Sun would appear to rise briefly, set, rise again, then proceed across the sky. At sunset it would set briefly, rise, then set again.

What Is Density?

Density is a measurement of how much mass is packed into a given volume of space. The density of all matter is compared to the density of water, which is 1.0, meaning 1 gram (mass) per cubic centimeter (volume). At 5.5, Earth's mean density is about five and a half times that of water and the highest of all the planets. The mean density of gaseous Saturn is the lowest at 0.7.

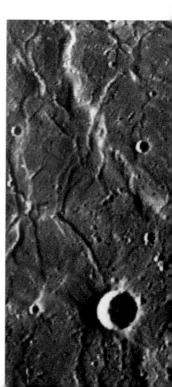

Temperatures and Seasons

From day to night, Mercury's surface temperature changes 1,130°F (630°C), more than any other planet or moon in the solar system. Just before sunrise on a typical day on Mercury the temperature is −300°F (−184°C). By midmorning the temperature rises to 80°F (27°C). At noontime, 22 Earth days since sunrise, it has climbed to 765°F (407°C). In the early afternoon the temperature reaches a high of 800°F (427°C), hot enough to melt zinc and tin.

On Earth, seasons change in a regular pattern due to the tilt of the rotational axis. Each hemisphere receives more nearly direct sunlight during one part of the orbit than in the other. Mercury's axis, in contrast, is very nearly perpendicular to its orbital plane, so no seasonal changes occur.

Mercury's Magnetic Field

It is thought that the magnetic fields of terrestrial planets are generated by convective motion in their molten metal cores. Since small planets lose heat more quickly than large ones, it was thought that Mercury's core might have completely solidified and thus become incapable of generating a magnetic field. Mercury's slow rotation also suggested the absence of a significant magnetic field. To the surprise of planetary scientists, however, a distinct magnetic field of about 1 percent of the Earth's was measured. The magnetic poles coincide with Mercury's rotational poles, as on Earth. Speculation that Mercury's field is a "fossil field," locked into an iron-bearing mantle when it solidified, could only be true if the temperature of the planet's mantle is below 932°F (500°C). Above this temperature, rocks lose their magnetism. Another theory suggests that Mercury's magnetic field is induced by the solar wind.

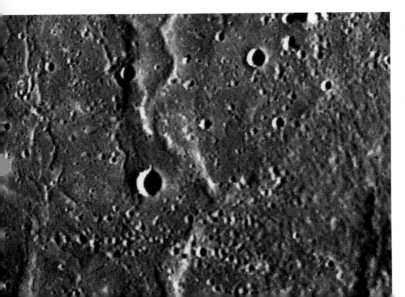

This close-up of the Caloris Basin shows the wrinkle ridge and fractures in the basin floor. The settling, due to gravity, causes the stretching and fracturing, and shrinking and folding of the basin's surface.

This global mosaic of Venus combines radar images acquired over two years by the *Magellan* spacecraft. Hues are based on color images returned by *Venera 13* and *14* landers. Bright areas reveal the equatorial highlands known as Aphrodite Terra.

Venus

Viewed by the naked eye, Venus is the brightest object in the sky after the Sun and the Moon. It orbits the Sun between Mercury and Earth, coming closer to Earth than any other planet. Venus is also close to Earth in size, with 95 percent of Earth's diameter and 80 percent of its mass.

Venus shines so brightly because its dense cloud cover reflects the Sun's light twice as well as the surfaces of the Moon or Mercury. Cloud coverage on Venus is 100 percent (compared to an average of 40 percent on Earth), so optical telescopes and spacecraft cameras are unable to reveal anything about the surface of Venus. The first sketchy views came from Earth-based

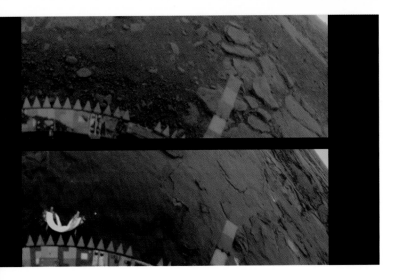

These images of the surface of Venus were returned by the *Venera 13* (top) and *Venera 14* (bottom) landers. The landscape and sky has an orange cast because the atmosphere filters out the blue component of sunlight.

radar images. Microwaves, or radar waves, invisible to the human eye, can penetrate clouds. Microwaves bounce off the high features and return to Earth before microwaves that bounce off lower features. Crude maps of the surface have been made based on this difference in arrival times. The best radar images are obtained when Venus is closest to Earth. But because the planet rotates exactly five times between closest approaches, the same face is always turned toward Earth.

Many people assumed that Venus was the "sister planet" of Earth because of their proximity and similarities in size and mass. But the robotic probes disproved this theory. They found that Venus, named for the Roman goddess of love and beauty, is one of the most hostile-to-life of the inner planets.

Exploring a Hostile World

Although it is twice as far from the Sun as Mercury, the surface of Venus is the hottest in the solar system. Its atmosphere is a dense layer of gases, mostly carbon dioxide, that traps the Sun's heat. Surface temperatures can rise to 900°F (480°C), hot enough to melt sulfur, lead, zinc, and tin. These conditions hampered exploration by spacecraft launched by both the former Soviet Union and the United States.

Soviet Missions Beginning in the 1960s, the Soviet Union launched a series of *Venera* missions to Venus. In October 1967, before it destructed during its descent, *Venera 4* was the first spacecraft to provide usable data on the planet's atmosphere, including its chemical composition, pressure, and temperature. *Venera 7* made the first transmission of data from the surface of another planet in December 1970. During the 1980s, landers of subsequent *Venera* missions sampled soil, provided more detailed atmospheric data, and transmitted color images of the surface,

Magellan Probe

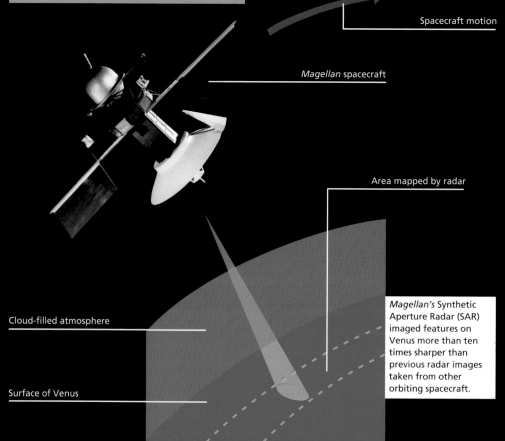

Spacecraft motion

Magellan spacecraft

Area mapped by radar

Cloud-filled atmosphere

Surface of Venus

Magellan's Synthetic Aperture Radar (SAR) imaged features on Venus more than ten times sharper than previous radar images taken from other orbiting spacecraft.

At 6:02 A.M. on October 12, 1994, the five-and-a-half-year *Magellan* mission concluded as ground controllers lost contact with the spacecraft. *Magellan* performed as planned on its final experiment, gathering data on the upper atmosphere of Venus even as it sank to its expected fiery doom. Most of the spacecraft was expected to burn up in the planet's atmosphere within a day, although some pieces of it probably reached the surface of Venus.

The *Magellan* spacecraft, named after the sixteenth-century Portuguese explorer who first sailed around the Earth, was launched May 4, 1989 and arrived at its destination—the planet Venus—on August 10, 1990.

Magellan's primary objective was to map the cloud-shrouded surface of Venus using radar. Besides mapping 98 percent of the surface to a resolution of 390 feet (120 m) or better, it also completed a gravity field map of 95 percent of Venus. This data will fuel scientific study for years.

Magellan also performed a first-of-its-kind "aerobraking" maneuver, dipping into the atmosphere of Venus to change its speed and reshape its orbit. This fuel-saving technique can be used in future missions.

including the only panoramic views of Venus we have. Orbiter missions *Venera 15* and *16* were the first spacecraft to obtain radar images of the surface from Venus orbit. In 1985 the atmospheric balloon probes and landers of the *Vega* project gathered even more information.

U.S. Missions *Mariner 2,* launched in 1962, was the first spacecraft to make a flyby of Venus. Passing within 21,000 miles (33,800 km) of the planet, it recorded high surface temperature and the lack of a significant magnetic field. As the Soviets' *Venera 4* probed the atmosphere, *Mariner 5* made a flyby and confirmed the high temperatures measured by *Mariner 2.* In 1974, before arriving at Mercury, the *Mariner 10* probe returned additional images and data on Venus.

In January 1979, the United States launched *Pioneer-Venus* into orbit around Venus. It made the first topographic survey of the planet using a radar altimeter to determine the elevation of surface features.

The United States sent the *Magellan* spacecraft into orbit around Venus in August 1990. Capable of revealing features not much larger than a football field, it has provided the most detailed view of the planet's surface and returned the highest-resolution radar images ever obtained from orbit. During some 15,000 orbits of Venus between August 1990 and October 1994, *Magellan* imaged about 98 percent of the planet's surface and greatly improved our knowledge of its gravity and topography. Its final experiment was conducted during its planned descent into the punishing Venusian atmosphere (see page 58).

The volcano Sapas Mons, named for a Phoenician goddess, sits in the center foreground of this computer-generated three-dimensional perspective view of the surface of Venus. The view combines images and elevation data returned by *Magellan.* The simulated color, based on images from Soviet *Venera* missions, was produced in 1992 by the Solar System Visualization Project and the *Magellan* Science Team.

240°   270°   300°   330°   0°   30°

Metis Regio

Ishtar Terra

60°   Lakshmi Planum   Maxwell M

50°

40°

30°

20°   Beta Regio

10°

0°

-10°

-20°   Phoebe Regio

-30°   Alpha Regio

-40°

Themis Regio   Lavinia Planitia
-50°

-60°

Lada

240°   270°   300°   330°   0°   30°

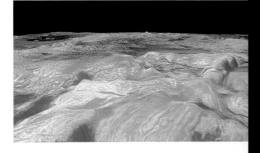

Magellan radar data was superimposed on topography to create this computerized perspective image of the highlands of Ovda Regio.

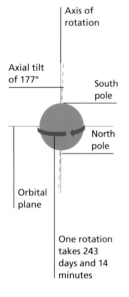

Axis of rotation

Axial tilt of 177°

South pole

North pole

Orbital plane

One rotation takes 243 days and 14 minutes

Orbit and Rotation

Venus orbits the Sun in about 225 Earth days. Its orbit is nearly circular, and it passes within about 25 million miles (40 million km) of Earth—closer than any other planet.

The axis of rotation for Venus is tipped nearly upside down when compared to the other inner planets. Therefore, it rotates in the opposite direction from Mercury, Earth, and Mars. Venus takes about 243 Earth days to complete one rotation, so its day is longer than its year. If you were standing on Venus, you would see the Sun rise in the west, not in the east, and you would have to wait about 61 Earth days for the Sun to reach high noon. It would finally set in the east after 121.5 Earth days.

Naming Features

Perhaps because Venus is the only planet named after a goddess (the Roman goddess of love), the International Astronomical Union, which selects names for all celestial objects and planetary features, decided to honor famous women in history by naming the features of the planet after them. The three exceptions are features that had been detected by Earth-based radar before that decision was made. They are highland areas named Beta Regio, Alpha Regio, and Maxwell Montes, the latter for James Clerk Maxwell. He was a noted mathematician and physicist of the 1800s, whose work on electromagnetism led to the development of radar.

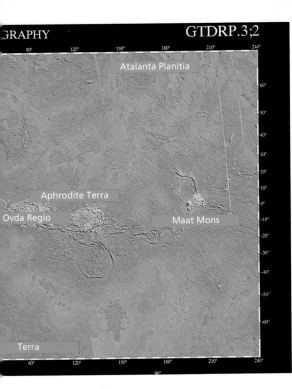

The topographical map of Venus was created by computer, using data from *Magellan*. Elevation is coded by color, with blue denoting the lowest elevations and red the highest. Artificial shading was added to highlight relief. This was created at the Massachusetts Institute of Technology's Center for Space Research.

Landscape of Venus

The terrains of Venus fall into three categories. Highlands make up less than 20 percent of the surface and rise higher than 1.2 miles (2 km) above the mean planetary radius, a standard altitude similar to sea level on Earth. Lowlands, also referred to as plains, lie below the mean planetary radius and make up about 40 percent of the planet's surface. The rest of the landscape is rolling plains, between 0 and 1.2 miles (2 km) above the mean planetary radius. In many areas they lie between the lowlands and highlands. The highlands of Venus can be compared to Earth's continents, and the lowlands to its ocean basins.

Large parts of the surface are flat—80 percent of Venus lies within a range of 0.6 miles (1 km) of the mean planetary radius. The total range of elevations, however, is similar to Earth's—about 8 miles (13 km), from lowest to highest.

Highland Regions The most predominant highlands of Venus lie in the northern hemisphere and around the equator. In the north, the landmass Ishtar Terra forms the high-latitude upland. Ishtar is larger than the continental United States. Lakshmi Planum, the plateau that forms the western part of Ishtar, is twice as large as the Tibetan plateau on Earth and rises higher. In the eastern part of Ishtar, the mountains of Maxwell Montes rise nearly 7 miles (11 km), higher than Mount Everest on Earth.

Aphrodite Terra dominates the equatorial highlands. A landmass comparable in size to South America, it stretches nearly halfway around the planet in the equatorial region. About 4 miles (6 km) at its highest, it is rougher and more complex than Ishtar Terra. Several canyons nearly 2 miles (3 km) deep, more than 100 miles (160 km) wide, and hundreds of miles long cut the eastern portion.

Beta Regio, a much smaller region, forms part of the equatorial highlands north and east of Aphrodite Terra. Its pair of shield volcanoes lie on either side of a rift valley. They rise about 3 miles (5 km) above the surrounding plains.

The Lowlands The lowlands are found predominantly in the mid-latitudes. One lowland area, Atalanta Planitia, has a roughly circular shape.

Volcanism

Venus was and still may be volcanically active. The indications are everywhere: large shield volcanoes like the ones on Hawaii, fields of smaller shield volcanoes, lava flows and channels, volcanic domes and collapsed volcanic craters known as *calderas,* and extensive volcanic plains.

Venus's Ovda Regio highlands was named for a volatile spirit from Finnish folklore. *Magellan's* images revealed a complex terrain consisting of ridges and troughs called *tesserae.* Volcanic or wind-blown material fills some of the troughs.

Volcanoes on Earth are concentrated in linear zones, but on Venus they are found throughout the rolling plains and in some highland areas. Major concentrations of volcanic features lie in an area bordered by Beta Regio, Alpha Regio, and Themis Regio. The lowlands, which appear to be dominated by flood volcanism, lack major volcanic features.

Venusian volcanoes tend to be broad but not as high as volcanoes on Earth. The highest volcano on Venus, Maat Mons, rises

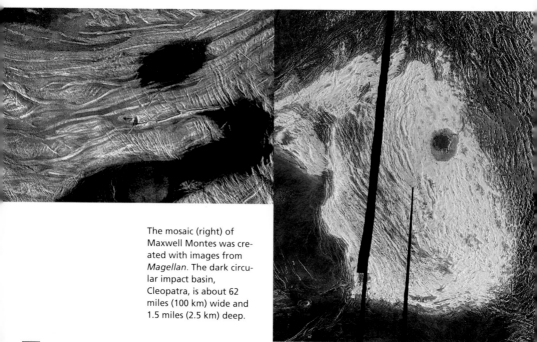

The mosaic (right) of Maxwell Montes was created with images from *Magellan*. The dark circular impact basin, Cleopatra, is about 62 miles (100 km) wide and 1.5 miles (2.5 km) deep.

5.3 miles (8.5 km) above the surrounding plains. By comparison, the Hawaiian volcanoes rise about 6.2 miles (10 km) above the sea floor. Some volcanic features on Venus have no counterparts on Earth. One such group is called the pancake domes, formed when thick or viscous lava oozed out onto the surface.

The *Magellan* spacecraft found hints, though no proof, of current volcanic activity. Its radar images and data show some rough lava flows comparable to young lava flows on Earth.

Volcanism reflects processes by which planets lose interior heat. Smaller bodies, like Mercury, radiate heat away through *conduction*. Larger bodies, like Venus, lose heat through hot spot volcanism (see advection). Hot spots produce shield volcanoes and flood volcanism and they have played a major role in forming the landscape of Venus.

Tectonics

As with volcanism, it appears that tectonic activity, the movement and resulting change in shape of a planet's crust, is widespread on Venus. Tectonically deformed regions known as *tesserae* dominate the highlands. The tesserae were probably formed by large amounts of crustal shortening, causing piling or stacking up of crustal blocks. Mountains, like Maxwell Montes, that are found on the margin of Ishtar Terra, also seem to have been formed by crustal shortening. Earthlike plate tectonics (lithospheric plates driven by flow in the mantle slowly collide forming mountains, volcanoes, and trenches) does not appear to have occurred on Venus in its recent geologic past.

The most common tectonic features on the planet, called *wrinkle ridges,* are found on the lowlands and rolling plains. Wrinkle ridges are found in volcanic plains on all the terrestrial planets.

The flat tops of the unique Venusian pancake domes (left) were cracked when the thick lava cooled and withdrew. The domes are less than 0.6 miles (1 km) high and about 39 miles (65 km) across. The *Magellan* radar image (below) shows Franklin, an impact crater in the Thetis region. The bright flows extend over 186 miles (300 km).

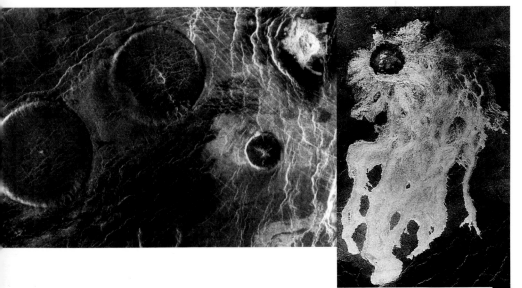

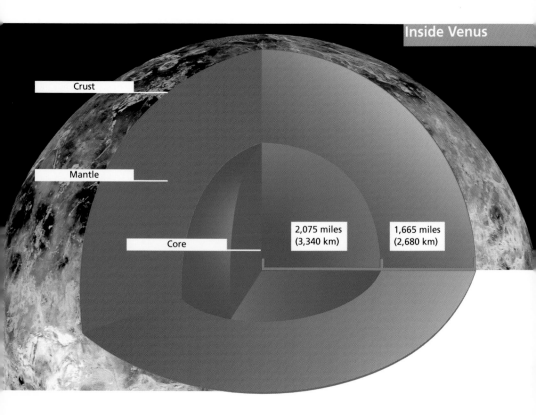

They occur when relatively small amounts of crustal shortening wrinkle or buckle the crust into ridges. On Venus, wrinkle ridges are up to 1 mile (2 km) wide and may be more than 100 miles (160 km) long.

Also common in the lowlands and plains are fractures that occur when the crust pulls apart or is stretched. These fractures have remarkably consistent patterns over very large areas. Rift valleys, caused by larger amounts of crustal stretching, are found in the rolling plains regions and are often associated with volcanoes. Ridge and fracture belts are elevated areas where moderate amounts of crustal shortening have buckled or fractured the planet's crust.

Volcano-Tectonic Features

Volcanism and tectonics sometimes combine on Venus, creating phenomena with no counterpart on Earth. *Arachnoids* resemble spiders whose "legs" are fractures formed by rising magma

Inside Venus

Crust

Mantle

Core

2,075 miles
(3,340 km)

1,665 miles
(2,680 km)

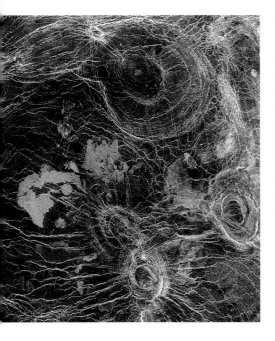

The Arachnoids

Prominent Venusian surface features called arachnoids resemble spiders. They were revealed by the spacecraft *Magellan*. So far observed only on Venus, arachnoids seem to be created when molten magma rises up from the interior and cracks the planet's crust. Volcanic activity in this region is also indicated by the presence of lava flows, which appear as bright patches in the center of this image.

and whose bodies are caldera-like collapse features. *Coronae* are bowl-shaped features surrounded by a depression or moat. They are believed to form when a bubble of hot magma in the mantle, called a *diapir*, pushes up the *lithosphere* and crust. The circular plateau that forms is fractured and covered by lava flows. Once the hot diapir cools, the interior of the plateau sinks.

Cratering

Venus was probably subjected to the same bombardment as Mercury, Earth, the Moon, and Mars, yet it has only about 900 impact craters on its surface—far fewer than one would expect. The smallest of these craters is about 2 miles (3 km) across. Smaller objects, which would make smaller craters, are destroyed by the planet's dense atmosphere before they reach the surface. Some Venusian craters have unique outflows. Their sources are in the crater *ejecta*, the material blown out in the explosion.

In general, older planetary surfaces accumulate more craters than younger ones. The average age of the surface of Venus is only 500 million years old. The oldest surface is less than a billion years old, relatively young by geologic standards. This could mean that many of the oldest terrains of Venus were covered by lava flows in a planet-wide volcanic event that began about 800 million years ago. It could also mean that the planet's lithospheric plate or plates were completely recycled (see page 75) at about this time.

Aurelia Crater, named after the mother of Julius Caesar, is a typical Venusian impact crater. When a meteoroid strikes the hot surface of the planet, both the meteoroid and the surrounding impact surface are vaporized. The resulting crater ejects material in a "flower petal" pattern, and lava trapped below the surface is released. Aurelia Crater is about 20 miles (30 km) across.

Atmosphere

The atmosphere on Venus is extremely dense, consisting of 96 percent carbon dioxide, 3 percent nitrogen, and only 0.5 percent water vapor. The planet's rotation is so slow that the atmosphere overheats. This creates winds in the Sun-facing hemisphere with speeds of 217 miles per hour (350 km/hr), three times the speed of hurricane-force winds on Earth.

Sulfur dioxide abounds in the lower atmosphere, probably because of volcanic activity. Hot sulfur dioxide and water vapor rise to the planet's upper atmosphere, where they are broken down by ultraviolet light. They recombine to form the planet's yellow clouds of concentrated sulfuric acid, which condense and fall as rain. Before the rain reaches the planet's surface, the extreme heat breaks down the sulfuric acid again into sulfur dioxide and water vapor.

The dense carbon dioxide in the atmosphere allows visible light to pass through to heat the surface but absorbs or blocks the flow of heat back into space. In this way, the atmosphere traps the heat radiated from the surface. This is known as the *greenhouse effect* because the carbon dioxide performs the same function as glass in a greenhouse. In moderation, it is a beneficial process. Mars and Earth are both warmer because of their greenhouse gases. But with an average surface temperature of 855°F (457°C), Venus is an extreme case.

The atmospheric pressure on the surface of Venus is 90 times greater than that on Earth—the equivalent of what a person would experience at about 3,000 feet (900 m) below the ocean surface, well beyond the range of scuba divers.

The Runaway Greenhouse Effect In the early solar system, the energy output of the Sun was about 25 percent less than it is today, so the Sun was dimmer. Venus may have looked a lot like Earth, with large bodies of water, even oceans, and an atmosphere consisting mainly of nitrogen. As the energy output of the Sun increased, Venus began to warm. Evaporation increased and more water vapor entered the atmosphere. Water vapor is an even more effective greenhouse agent than carbon dioxide, so the warming increased further. Carbon dioxide that had dissolved in the oceans was released into the atmosphere by the greater evaporation. The increased heat also released carbon dioxide that was stored in carbonate rocks such as limestone. When the temperature rose to the boiling point of water, the oceans would have slowly boiled away.

Why then is there so little water in Venus's atmosphere today? As the amount of water vapor in the lower atmosphere increased, more and more water vapor made its way to the upper atmosphere. Here it was broken down into its components—hydrogen and oxygen—by ultraviolet light. The free hydrogen then escaped into space. This is referred to as the runaway greenhouse effect.

Both these images show the swirling sulfuric clouds of Venus's upper atmosphere as recorded by the *Pioneer-Venus* spacecraft, which arrived at Venus in December 1978. The fast-moving clouds move from east to west across the planet.

If you were there

If you weighed 100 pounds (45 kg) on Earth, you would weigh about 91 pounds (41 kg) on Venus.

On the surface, you would find the atmospheric pressure crushing and the temperature unbearable.

Few have had the unique experience of gazing at the startling beauty of their home planet from outer space. This image was taken during the *Apollo 10* mission to the Moon in May 1969.

Earth

The third planet from the Sun is one of the most geologically active planets in the solar system, with large volcanoes and great mountain chains. Water is abundant, as vapor in the atmosphere and as liquid and ice on the surface. Earth is the only planet in the solar system with an average temperature between the freezing and boiling points of water. Its most unique characteristic is that it supports an amazing diversity of life.

Early History

The forces that shaped Earth's landscape in the past—tectonism, volcanism, plate recycling, and erosion by wind, water, and glaciers—are constantly working to reshape the surface today. Most of Earth's surface, therefore, is relatively young, less than 100 million years old. Although Earth is about 4.5 billion years old, the oldest known terrains are only 3.5 to 3.7 billion years old. No terrain from Earth's earliest history survives, but our planet probably began much like the other terrestrial planets.

Earth was formed about 4.5 billion years ago by the accretion of rocky masses. As this mass increased, the temperature rose dramatically and the rocky material melted and separated. Eventually the dense, heavy material, mostly iron and nickel, sank to form the central core while less dense, lighter, mostly silicate material rose to the surface to form the planet's crust.

As Earth cooled, it was bombarded by objects from space. Perhaps the largest of these objects was the size of Mars, blasting a cloud of debris into orbit that would eventually become the Moon. Before the bombardment ended, about 3.8 billion years ago, 35 impact basins larger than 186 miles (300 km) across were formed on the Moon. The larger and more massive Earth probably experienced hundreds of comparable impacts. Some would have formed gigantic impact basins with diameters equal to half the width of the continental United States.

Water

Earth is the only inner planet that has liquid water on its surface, although Mars and Venus may have had surface water at one time. As Earth cooled, a primitive atmosphere was formed through a volcanic process known as *outgassing*. This is the release of gases from the interior, including water vapor, hydrogen, nitrogen, and carbon dioxide. As the water vapor in the atmosphere increased, it condensed and fell as rain. Eventually enough water collected to form our oceans.

Oceans cover 70 percent of Earth's surface. If they were spread out evenly, Earth would be under 2.5 miles (4 km) of water. More than 80 percent of all *photosynthesis* (the process by which plants convert sunlight into chemical energy) takes place in the oceans, making them the principal habitat of life on Earth.

Ocean water contains many minerals leached, or dissolved out, from rocks and soil and carried to the oceans by rivers. Sodium chloride, or common table salt, composes 3.5 percent of ocean water.

Fresh water, containing almost no salt, is essential for most living things not found in the oceans. The source of fresh water is

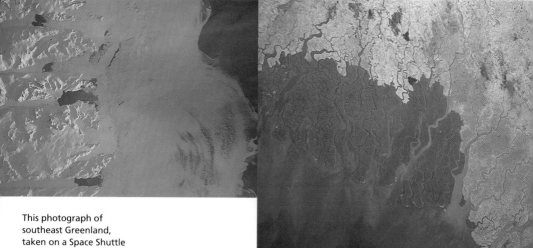

This photograph of southeast Greenland, taken on a Space Shuttle mission, shows coastal fjords partially filled by glaciers projecting into the North Atlantic Ocean.

This color-enhanced image of the shifting channels of India's sacred Ganges River was returned by Space Shuttle *Atlantis* in November, 1985. Surrounding vegetation is indicated in red.

rain, which comes from pure water vapor evaporated from the oceans. The proportion of fresh water to ocean water is small. Only a little over 2 percent of the total amount of water on or near Earth's surface is fresh water. Of this fresh water, about 80 percent of it is frozen in glaciers at the poles.

Atmosphere

Much of Earth's atmosphere and water were probably outgassed early in the planet's history. Carbon dioxide, nitrogen, hydrogen, water vapor, and other *volatile* elements (that vaporize at relatively low temperatures) comprised Earth's early atmosphere. These elements may have come both from the rocky material on Earth and from the volatile-rich material from the outer solar system.

For the first billion years, oxygen was only a small component of Earth's atmosphere. Oxygen accumulated in the atmosphere only after photosynthesis began producing it faster than it was lost by chemical combination with other gases and metals. Most of the carbon dioxide in Earth's early atmosphere was removed not by plant life but by chemical combinations with calcium, hydrogen, and oxygen, which form limestone (calcium carbonate) in the oceans.

In recent times humans have been generating carbon dioxide faster than plants or the oceans can take it up. At the same time,

Winds blow the plume of the Mount Etna volcano eastward from the island of Sicily over the Ionian Sea. This photo, taken during the July 1975 *Apollo-Soyuz* mission, shows the toe of Italy in the foreground and Sicily across the narrow Strait of Messina.

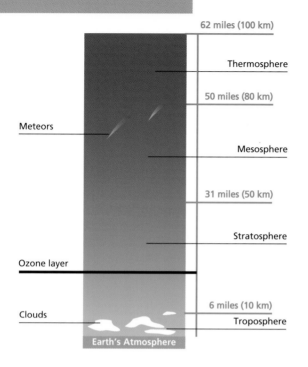

62 miles (100 km)

Thermosphere

50 miles (80 km)

Meteors

Mesosphere

31 miles (50 km)

Stratosphere

Ozone layer

6 miles (10 km)

Clouds

Troposphere

Earth's Atmosphere

we are cutting down forests that use carbon dioxide and burning the wood, releasing more carbon dioxide. Earth's atmosphere is now about 78 percent nitrogen, 20 percent oxygen, 1 percent argon, and 1 percent water vapor. The average surface pressure is about 15 pounds per square inch (1 kg/sq cm).

Seasons

Seasonal weather changes on Earth are largely a matter of geometry. Earth follows a nearly circular orbit. The tiny difference of 3 percent between its greatest and smallest distance from the Sun does not account for the range of temperatures between the seasons. The explanation lies in Earth's axis, which is tipped 23.5° off the perpendicular to the plane of its orbit. When the northern hemisphere tips toward the Sun, the north pole is continuously lighted and days are longer everywhere north of the equator. There is also more heat. The Sun's rays reach the northern hemisphere at less of an angle, with less filtering by the atmosphere, so the surface absorbs more heat each hour. For part of the year more heat is gained each day than is lost at night through radiation.

Seasons change each time Earth moves a quarter of the way around its orbit. The longest day, on about June 21, is called the *summer solstice* in the northern hemisphere. Three months later, on about September 23, day and night are the same length. This is the *autumnal equinox*. The shortest day of the year, the *winter*

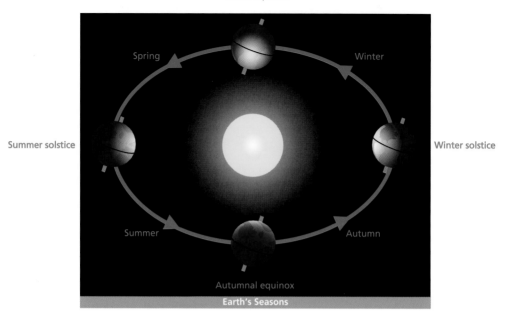

Earth's Seasons

solstice, occurs on about December 22, when the southern hemisphere is tipped toward the Sun. At this time the north pole is always dark and the northern hemisphere is losing more heat each night than it gains during the day. At the *vernal equinox,* on about March 21, nights and days are again the same length.

Earth's Structure

The interior of Earth has three distinct parts: the core, the mantle, and the outer shell, which includes the lithosphere and crust. Like the interiors of the other terrestrial planets, it was shaped by a process known as *differentiation,* the separation and migration of materials according to their relative densities. Whether these materials exist as gases, liquids, or solids depends on both temperature and pressure.

The Core Earth's core has an *inner core* and an *outer core.* The inner core makes up only 1.7 percent of Earth's total mass and is solid iron and nickel. It is extremely hot but remains solid because the weight of the overlying material exerts enormous pressure on it. At the center of Earth the pressure is about 3.7 million times greater than the pressure at the surface.

The solid inner core is surrounded by a liquid outer core that accounts for 30.8 percent of Earth's mass. The temperature of the liquid core is about 7,400° F (4,100° C). The outer core is liquid, rather than solid, because as distance from Earth's center increases, pressure decreases. Its density is about 11. Convective currents in the liquid outer core are thought to produce electrical currents that generate Earth's strong magnetic field.

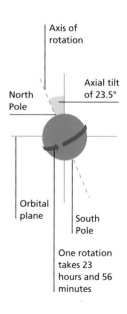

Axis of rotation

Axial tilt of 23.5°

North Pole

Orbital plane

South Pole

One rotation takes 23 hours and 56 minutes

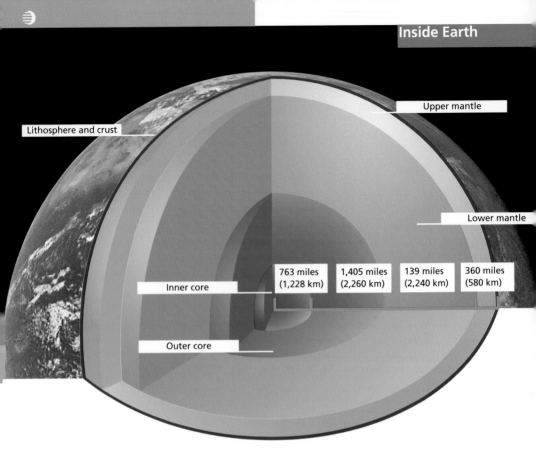

Upper mantle

Lithosphere and crust

Lower mantle

| 763 miles | 1,405 miles | 139 miles | 360 miles |
| (1,228 km) | (2,260 km) | (2,240 km) | (580 km) |

Inner core

Outer core

The Mantle The liquid outer core is surrounded by a layer called the *mantle*, which makes up 67 percent of Earth's mass. The mantle extends from a depth of about 43 miles (70 km) to about 1,800 miles (2,890 km). Most of the mantle is solid, although the upper mantle appears to be partially molten in a region called the *asthenosphere*, about 43 to 250 miles deep (70 to 400 km). Below the asthenosphere, the mantle is hotter but more solid because of the increased pressure.

The mantle is thought to consist of rocks composed of the minerals pyroxene, olivine, and garnet. The partial melting of the upper mantle produces a liquid with the composition of *basalt*— a common volcanic rock that makes up the sea floor and shield volcanoes, like those in the Hawaiian islands.

The Lithosphere and Crust Resting on the mantle is the outer shell, which is divided into two parts: the relatively cold, rigid *lithosphere* and the *crust*. This shell accounts for about 0.4 percent of Earth's mass. The crust, the outermost part of the shell, is proportionately as thin as the skin of an apple in relation to the diameter of Earth. The continents are embedded in the lithosphere, a transition zone between the hot mantle and the cold crust, with an average thickness of about 43 miles (70 km). The lithosphere is not a single continuous shell but a mosaic of tightly fitting plates—eight large ones and a few dozen smaller ones.

The lithosphere plays a key role in a process called *plate recycling*. Individual plates are constantly moving, being pushed back into the mantle and destroyed at some plate boundaries—a process called *subduction*—and being formed again at other boundaries. The slow movement of the plates is driven by convection currents in the mantle.

Plate recycling begins on the ocean floor, where new lithosphere is being formed at *midocean ridges*. These ridges occur on most sea floors and reach heights of about a mile (1.6 km). In the transition zone between the upper and lower mantle, material upwelling from deep in the mantle causes partial melting, and magma erupts at the midocean ridges to form new sea floor, or oceanic crust. As more magma erupts and pushes out, the new sea floor spreads away from the ridge, cooling and thickening. This process is called *sea floor spreading*.

As the new sea floor gradually cools and moves farther from the spreading ridge, it becomes thicker and denser and tries to sink back into the mantle. Eventually, this now mature oceanic lithosphere reaches continental lithosphere, which is much older and thicker—about 93 miles (150 km). The thinner, denser oceanic lithosphere is forced down underneath the continental lithosphere and back into the mantle along *subduction zones*. Deep ocean trenches are associated with these zones.

Plate recycling keeps the sea floor young. The floor of the Atlantic Ocean, for example, is spreading at the rate of almost half an inch (1 cm) per year. The average age of the sea floor is only 60 million years; nowhere is it older than 200 million years.

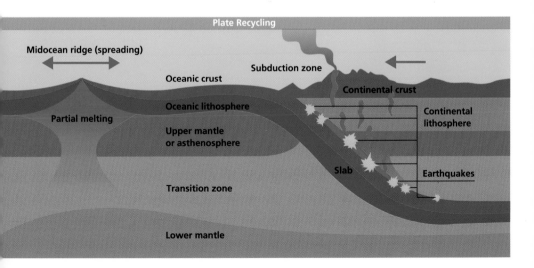

Plate Recycling

Midocean ridge (spreading)

Subduction zone

Oceanic crust

Continental crust

Oceanic lithosphere

Continental lithosphere

Partial melting

Upper mantle or asthenosphere

Slab

Earthquakes

Transition zone

Lower mantle

Cratering, Volcanism, and Tectonics

The processes called cratering, volcanism, and plate tectonics create changes in the surfaces of Earth and other bodies in the solar system. Some of the basic dynamics of each process are shown here, although these dynamics vary on individual planets and satellites.

Collision path of meteorite

Ejecta (impact debris)

Wall of rock surrounding crater

Secondary crater

Saucer-shaped crater

Fractured rock

Cratering

Impact craters are formed when an object strikes the surface of a planet or satellite. When the object hits the surface the impact energy fragments, melts and ejects material forming a crater cavity. Pieces of debris may create secondary craters around the major impact site. Craters on the Moon and the inner planets are created this way. On Earth, however, most craters have been lost to erosion.

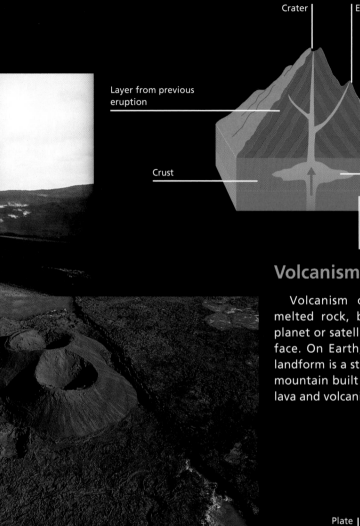

Crater | Eruption at side vent

Layer from previous eruption

Crust

Magma chamber

Lithosphere

Volcanism

Volcanism occurs when magma, or melted rock, beneath the surface of a planet or satellite breaks through the surface. On Earth, one example of volcanic landform is a stratovolcano, a cone-shaped mountain built up by alternating layers of lava and volcanic ash.

Tectonics

Tectonics refers to the movement and deformation of the crust of a planet or satellite. On Earth, deformation occurs when segments of Earth's lithosphere, called plates, interact in a process known as Plate Tectonics—as shown in this diagram. Mars, in contrast, is thought to have only one plate. Tectonic features are distributed across its surface, rather than concentrated at the margins of plates as on Earth.

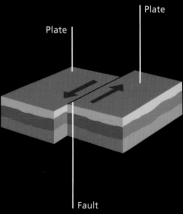

Plate

Plate

Fault

Plate Tectonics

The interaction among the plates of the lithosphere is called *plate tectonics*. Much of Earth's landscape has been shaped by plate tectonics. As oceanic plates have been created and subducted and the continents have collided and broken apart, mountains have been built, rift valleys formed, ocean ridges and trenches created, and volcanoes constructed.

Most activity occurs along the edges of the plates as they move in relation to one another. Plate tectonics begins with plate recycling and is driven by convective cooling of Earth's interior. The continental lithosphere is made largely of rocks such as granite that are less dense than the mantle. So, while oceanic plates are forced back into the mantle, the continents stay afloat and remain at the surface, although they have drifted together and broken apart many times. By 80 million years ago, most of the continents we know today were isolated and had begun moving toward their current positions.

The term "continental collision" conjures up visions of high-speed crashes, but the continents move only 2 to 4 inches (5 to 10 cm) per year, and it takes millions of years to build a mountain range. Around 250 million years ago, when North America collided with Africa, the ensuing large-scale crustal shortening generated the Appalachian Mountains. Erosion has worn them down, but they were once comparable to the present-day Himalayas. The Himalayas, however, were formed only about 35 million years ago when India plowed into southern Asia. The large-scale horizontal shortening that resulted built up the highest mountains on Earth.

Mount Everest (above), a relatively young mountain, towers more than 29,000 feet (8,800 m) above sea level on the border of Nepal and Tibet. The Great Smoky Mountains (below), part of the Appalachian chain in the eastern United States, show the effects of very long-term erosion. The tallest peak in this range is 6,643 feet (2,025 m) high.

The result of the collision of the Indian-Australian Plate and the Eurasian Plate, the Himalaya mountain system took millions of years to form. Consisting of several parallel ranges, parts of the system are 200 miles (320 km) wide. The Himalayas form a natural barrier between China and India and Nepal.

Earthquakes Stresses build when the plates come in contact, and only so much stress can be supported before the rock breaks, causing an earthquake. Earthquakes release enormous amounts of energy that travel through the Earth and along its surface as waves. Much of what we know about Earth's interior we have learned by studying how these waves travel through Earth.

Earth's plates not only collide with one another, they also can slip past one another. This is the case with the Pacific plate (see page 80) and the North American plate. The result is the infamous San Andreas fault in California. This type of contact produces shallow earthquakes that can be very destructive.

Deep earthquakes are associated with the subduction of oceanic plates. Some have been detected at depths exceeding 370 miles (600 km), indicating how far the plates may plunge into the mantle.

The San Andreas fault (right), slices through California for more than 700 miles (1,100 km). Fault activity accounts for more than 10,000 earthquakes—most of them minor—each year.

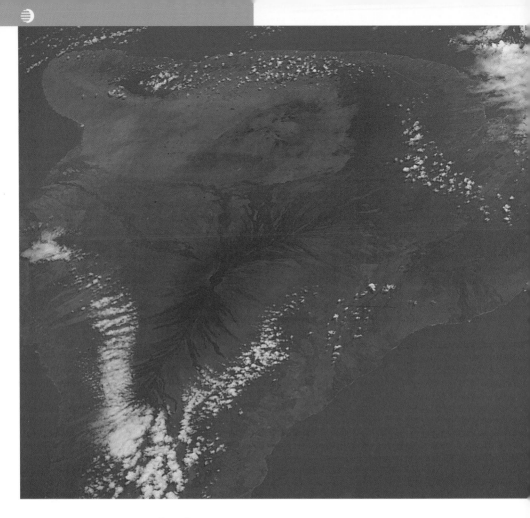

Volcanism

Volcanic activity on all four terrestrial planets reflects ways in which heat is lost from the interior. On Earth, there are three major types of volcanic activity. *Ocean-ridge volcanism* creates new oceanic lithosphere—the first step in the process of plate recycling described earlier. *Convergence-zone volcanism* produces composite or stratovolcanoes on plate margins. *Intraplate volcanism* produces shield volcanoes and flood basalt provinces in the interior of plates.

In convergence-zone volcanism, very steep stratovolcanoes are formed when magma rises through the continental crust. The lava usually contains large amounts of gases that are often released explosively, forming ash. Alternating eruptions of lava and ash build a cone-shaped mountain with steep slopes. Examples are Mount Fuji in Japan; Mounts Vesuvius, Etna, and Stromboli in Italy; and Mount Rainier and Mount St. Helens in the United States.

Hawaii's Kilauea volcano, on the Island's eastern coast, is currently experiencing a period of eruptions that began in 1983. Halemaumau Crater, inside Kilauea's caldera, held a lake of molten lava for about a century before it sank in a flurry of steam explosions in 1924.

In intraplate volcanism, magma comes from hot spots in the mantle. The basalt lava has only small amounts of gas and produces little ash. The relatively free-flowing lava spreads widely, building a broad, domed mountain with gentle slopes. The volcanoes on the island of Hawaii are examples. Mauna Loa rises 5.7 miles (9.1 km) above the sea floor and has a basal diameter of 60 miles (100 km). Because the Pacific plate on which Mauna Loa is built is in motion, it has moved over a stationary mantle hot spot. This has resulted in the chain of volcanoes that make up the Hawaiian Islands, each of which was once a vent for the same hot spot. If the Pacific plate had remained in the same position, a single enormous volcano would have formed over that hot spot rather than many smaller ones.

Hot spots are also responsible for the flood basalt areas on the continents. Magma from hot spots fractured the continental crust and rose through it. High volumes of free-flowing lava erupted through the cracks, or *fissures*, covering broad areas of the crust. The Columbia Plateau in the northwest United States is a good example. The flows there cover an area of 72,500 square miles (188,000 square km)—larger than the surface area of the state of Washington—and the volcanic pile reaches a thickness of 2.5 miles (4 km). Another, even larger flood basalt area is the Deccan Plateau in India.

The distribution of the 500 to 600 active volcanoes on Earth is not uniform on the plates. Most are found on plate boundaries where lithosphere is being created or destroyed. Many are on the margin of the Pacific plate, forming what has been called the Ring of Fire. Intraplate volcanoes account for only a small number of the active volcanoes. Thus the dominant heat-loss mechanism on Earth is clearly plate recycling.

Mount St. Helens, in Washington State, is shown before and after the 1980 eruption that blew its top off. The crater left by the blast diminished the volcano's height by more than 1,000 feet (300 m) and blanketed the surrounding area with volcanic ash.

The Manicouagan impact structure in Quebec, Canada is about 40 miles (65 km) across. The impact that formed it 210 million years ago melted the rock in the central plateau region. A frozen lake, which appears white in this Space Shuttle photograph, surrounds the plateau.

Impact Cratering

Of all the terrain on Earth that underwent the period of heavy bombardment, none has survived the effects of erosion, plate re-cycling, and plate tectonics. But, there is evidence that Earth has been hit by asteroids and comets, some recent. It is estimated that in the last 100 million years Earth has been hit by well over a thousand objects capable of producing craters with diameters ranging from 6 to 93 miles (10 to 150 km). About a hundred known impact structures on Earth range in diameter from 2 to over 93 miles (3 to over 150 km) and many others are thought to be impact features. Most are found on the oldest parts of the continents, called the *shields*. Only a few of the largest impact structures are older than 500 million years.

One well-preserved impact structure is Meteor Crater in Arizona, also known as Barringer Crater. It was formed about 50,000 years ago when an iron-nickel asteroid about 160 feet (50 m) wide hit flat-lying sedimentary rocks and blasted out a crater about 0.7 miles (1.2 km) wide and about 1,300 feet (400 m) deep. The impact was roughly equivalent to a 4-megaton nuclear explosion.

The Lonar Crater, located on the flood basalt areas of India's Deccan Plateau, is much smaller and younger than the Manicouagan Crater. It is 1.1 miles (1.8 km) across and about 50,000 years old.

The largest impact structure discovered to date is the Chicxulub Crater. It is well hidden under sediments on the coast of the Yucatan Peninsula. The crater was detected when instruments measuring variations in Earth's gravitational and magnetic fields showed a circular structure about 112 miles (180 km) wide and possibly a larger concentric structure about 186 miles (300 km) wide. The object that made the crater is estimated to have been about 6 miles (10 km) across. If the larger surrounding structure is the crater rim, the impact would have been roughly equivalent to 50 trillion tons (45 trillion MT) of TNT.

Not all impact events on Earth occurred in the distant past. In 1908 a stony body about 160 feet (50 m) in diameter, possibly the nucleus of a comet, penetrated Earth's atmosphere above Siberia. It may have been moving at a speed of 70,000 miles per hour (110,000 km/hr) before it disintegrated. The object left no crater, but it created a powerful shock wave in the atmosphere that flattened trees over a radius of nearly 20 miles (32 km). A phenomenon like this one, referred to as "the Tunguska Event," is likely to occur only about once every thousand years.

Earth: Plates, Volcanoes, and Earthquakes

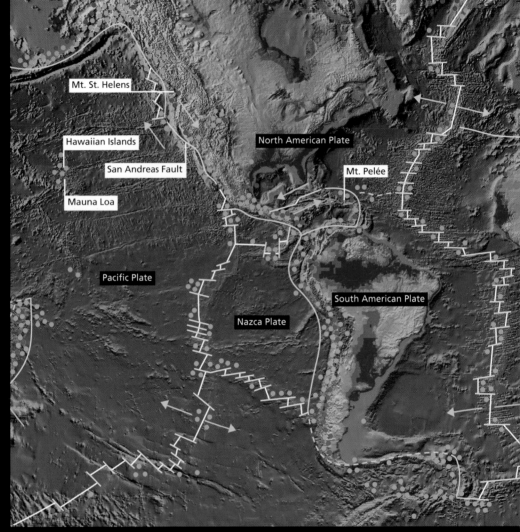

Mt. St. Helens

Hawaiian Islands

San Andreas Fault

Mauna Loa

North American Plate

Mt. Pelée

Pacific Plate

South American Plate

Nazca Plate

——— divergent boundary

——— transform fault

——— convergent boundary

- - - uncertain

➡ plate movement

● earthquake location

● volcano location

Most of the volcanoes, mountain systems, and earthquake belts are located on the edges or margins of Earth's lithospheric plates. The stresses that build up from the movement of these plates are the cause of powerful earthquakes. In zones where plates converge,

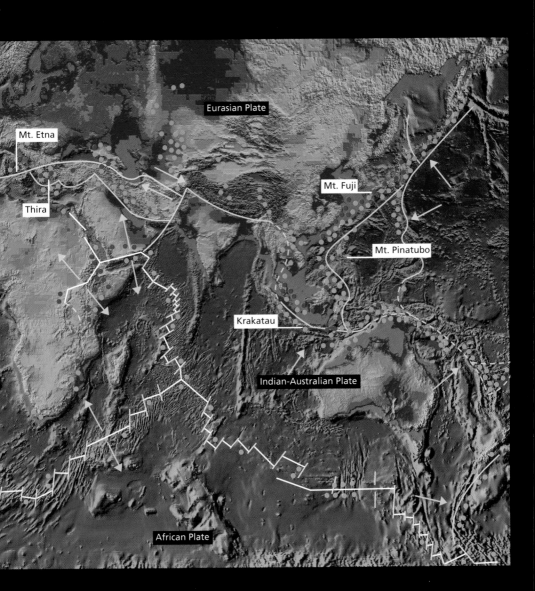

Eurasian Plate

Mt. Etna

Thira

Mt. Fuji

Mt. Pinatubo

Krakatau

Indian-Australian Plate

African Plate

mountains chains are pushed up, magma rises forming volcanoes, and deep-sea trenches are created. Some of Earth's volcanoes, however, are found on the interior of the plates away from their margins. They form over stationary hot spots in the mantle. The Hawaiian Islands are one example.

Life on Earth

There is evidence that life developed very early on Earth. Fossils of primitive cells have been found in rocks in South Africa and Australia that are 3.5 billion years old. But organisms large enough to leave clear fossil evidence did not evolve until about 600 million years ago, considered recent in geologic time.

Earth's atmosphere supplies the oxygen that animals, birds, fish, and humans require. Oxygen also reacts with sunlight to form ozone, which blocks out part of the life-destroying ultraviolet rays. Oxygen and nitrogen block out most of the rest. Small meteors strike our atmosphere constantly but are burned up by atmospheric friction as they fall. Our magnetic field protects life from high-energy particles ejected from the Sun and other sources in our galaxy.

Most of the Sun's energy falls on equatorial regions, but air and ocean currents convey vast amounts of warmth to other parts of Earth. The warming greenhouse effect from carbon dioxide in the atmosphere allows infrared energy to reach the surface but inhibits the radiation of heat back into space. Also, Earth's huge store of surface water helps stabilize the temperature, giving off warmth when the air is cold and absorbing heat when the air is hot.

From towering mountains to lush lowland forests, Earth's many terrains and climates support an endless variety of life forms.

The relative stability and mildness of Earth's climate have made the planet conducive to life. In the past, however, major climate changes were driven by changes in Earth's orbit, its axis, and the positions of the continents. Variations in Earth's orbital

The Ozone Layer

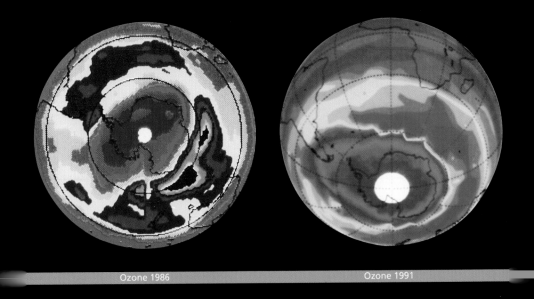

Ozone 1986   Ozone 1991

Through a series of chemical reactions involving sunlight, oxygen in the atmosphere forms ozone, which consists of three oxygen atoms (O_3). The greatest concentration of ozone is in the stratosphere at an altitude of about 19 miles (30 km). This ozone layer absorbs harmful ultraviolet radiation, which is dangerous to all life on Earth.

In the 1970s there was a growing concern that pollution might damage the ozone layer. Data on the concentration of ozone above the southern hemisphere, collected by polar-orbiting meteorological spacecraft, indicated that ozone levels over Antarctica had decreased by 40 percent between 1977 and 1984. Industrial compounds called chlorofluorocarbons (CFCs) are thought to have contributed to this decline. Ultraviolet radiation breaks down CFCs, releasing chlorine. This chlorine in

The size of the ozone hole—the area over Antarctica that displays intense depletion each year between late August and early October—reached a plateau during the winters of 1992, 1993, and 1994. It extended to about 9 million square mile (23,300 sq km) each year.

The Space Shuttle *Atlantis* completed a mission to collect data about the ozone hole in November 1994. *Atlantis* has conducted the Atmospheric Trace Molecule Spectroscopy (ATMOS) experiment, which showed that the recovery of the ozone layer during warmer months happens differently at different levels of the stratosphere. Germany's Cryogenic Infrared Spectrometers and Telescopes for the Atmosphere (CRISTA) experiment recorded levels of gases over the entire globe. The data collected from these and other missions will enhance our understanding

eccentricity (how much an orbit deviates from a circle) and in the *obliquity* of its axis (tilt of the axis with respect to the plane of orbit) vary on time scales of 10,000 and 100,000 years, respectively. Changes in the orientation of Earth's axis, called *precession,* occur over a 25,800-year period.

One of the most dramatic effects of these changes is ice ages. Periodic ice ages, during which the polar ice caps advanced halfway to the equator, occurred over the past million years and dramatically reshaped landscapes.

Climate changes caused by Earth's orbital and axial characteristics have also influenced the formation of deserts. Climate models that adjust for the precession of the axis suggest that about 10,000 years ago the Sahara Desert of Africa had ample rainfall, rich vegetation, and monsoons. But there may be other influences. While Earth's climate has been stable since the last ice age 10,000 years ago, analysis of ice core samples from Greenland shows that some earlier abrupt changes in climate lasted only decades or centuries. At one time, the average temperature plunged 25°F (14°C) in a decade, and the cold snap lasted for 70 years.

Challenges to Life The fossil record testifies to mass extinctions of species in the oceans and on land. Four global mass extinctions occurred about 11 million, 35 million, 66 million, and 91 million years ago, correlating with sedimentary layers that contain the element iridium, which is 10,000 times more abundant in most meteorites than in Earth's crust. It is thought that

Earth's varied terrain and its oceans are constantly changing, owing to the forces of nature and the effects of human activity on a wide range of environments.

the iridium came from volcanic eruptions deep in the mantle or from the impact of an asteroid. Other evidence suggests that large impacts occurred 35 million and 66 million years ago, throwing enormous amounts of dust into the atmosphere, blocking out much of the sunlight for a few months or even years, and possibly igniting great forest fires. In the mass extinction about 66 million years ago, 75 percent of the existing plant and animal species disappeared in less than a few million years, ending the age of the dinosaurs and ushering in the age of the mammals. Some believe this extinction was triggered by the impact that formed the Chicxulub Crater. Others point to the outgassing that accompanied the formation of the Deccan Plateau about 66 million years ago as the cause of the mass extinction.

Clementine, a small U.S. spacecraft sent to map the surface of the Moon in early 1994, captured this image of Earth from its lunar orbit later that year.

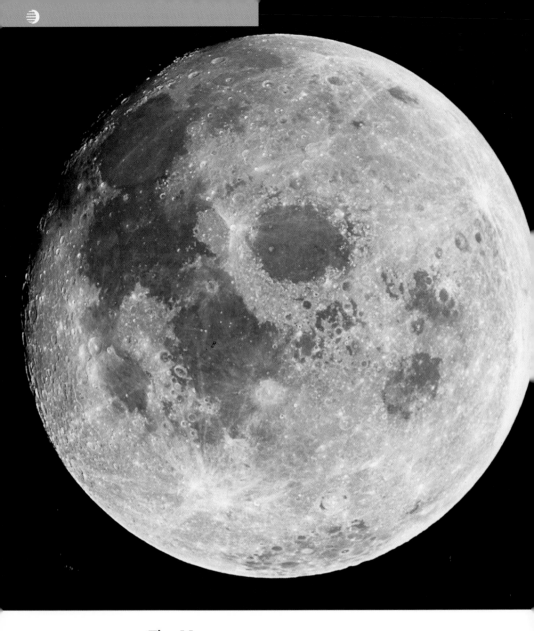

The Moon

Apollo 11 cameras took this picture of the full Moon in July 1969, 10,000 nautical miles (18,500 km) into its journey home.

Much of our interest in the solar system is rooted in our continued fascination with the Moon. Its presence in the night sky has captured human imagination throughout history. We have studied it with the naked eye, with telescopes, with orbiting spacecraft and landers, and, most revealingly, with human exploration of its surface. To date, the Moon has the distinction of being the only extraterrestrial object in the solar system to have been visited by humans.

NASA's *Apollo 11* mission achieved the first piloted lunar landing in 1969. Astronaut Edwin "Buzz" Aldrin is shown descending to the Moon's Sea of Tranquility.

Exploring the Moon

In the 1960s the United States and the Soviet Union launched a series of orbiters and landers with the mission of exploring the Moon's surface. These devices sent images and rock samples back to Earth.

Then, in one of the greatest of human achievements, astronauts on the *Apollo 11* mission stepped onto the Moon on July 20, 1969. Five subsequent expeditions landed a total of 12 U.S. astronauts to survey the lunar surface and retrieve rock samples. The nearly one-third of a ton (381 kg) of lunar rocks they returned to Earth are still being studied, and have told us most of what we know about the composition of the Moon. The astronauts also set up seismic stations to monitor "moonquakes" and study the lunar interior. The thousands of photographs taken from orbit and from the surface continue to be analyzed.

Exploration of the Moon has continued with the *Galileo* and *Clementine* spacecraft. In October 1989, *Galileo* began a six-year journey to Jupiter. On its way, it passed by the Moon twice and returned data on the composition of the lunar surface. *Clementine* went into lunar orbit in February 1994 and surveyed the Moon for several months. Equipped with laser-ranging system and high-resolution imaging cameras, *Clementine* imaged the entire surface and determined the topography to within 130 feet (40 m). In all, it returned about a million and a half detailed images that have been pieced together to create global mosaics and it will allow the composition of the surface to be mapped in detail.

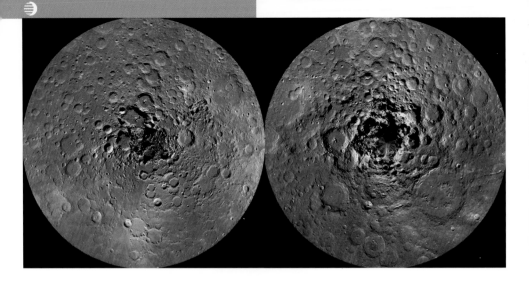

These mosaics of the lunar north (left) and south polar regions were made from images returned by the *Clementine* probe in early 1994. Prior to the *Clementine* mission, the topography of the Moon was thought to have a range in elevation of about 6 miles (10 km). Data returned by *Clementine* indicates a range of about 10 miles (16 km).

NASA's *Apollo 15* Lunar Module (LM) reached the Moon on July 31, 1971. After landing on the Hadley Rille region, the crew completed the first operation of a lunar roving vehicle (LRV).

Basic Characteristics

About a quarter the size of Earth, the Moon has only slightly more than 1 percent of Earth's mass. This is because the mean density of the Moon is only 3.3 compared to Earth's 5.5.

The Moon travels around Earth in an oval orbit at about 22,870 miles (36,800 km) per hour. The Moon's orbital distance from Earth varies from about 221,000 miles to 253,000 miles (356,000 km to 407,000 km). The Moon's surface gravity is about one-sixth that of Earth, so a 100-pound (45 kg) person would weigh about 17 pounds (8 kg) on the Moon.

Like Mercury, the Moon has no real atmosphere. Compared to Earth, it is relatively poor in volatiles—elements that vaporize at low temperatures—so it is unlikely that much of an atmosphere was outgassed during its early history. Even if the Moon had been rich in volatiles, its gravitational pull is too weak to hold on to an atmosphere. The Moon's *escape velocity*—the velocity required to carry a particle away from the Moon—is a mere 1.5 miles (2.4 km) per second.

Lacking the modifying effects of an atmosphere like Earth's, the Moon has extreme temperatures, ranging from −300°F (−184°C) at night to 214°F (101°C) at noon. At the poles, however, the temperature is a constant −140°F (−96°C). Some areas at the south pole may be in permanent shadow, staying cold enough to trap water in the form of ice from passing comets.

Though it looks perfectly round from Earth, the Moon is a little lopsided. The lunar crust is thicker on the far side than on the near side, which puts the Moon's center of mass 1.2 miles (2 km) closer to Earth than its geometric center.

The Lunar Landscape

The two major terrain types on the Moon are easy to identify with the naked eye. They form the features that make "the man in the Moon." The younger, dark plains are called *maria,* and the bright areas are highlands, or *terrae.*

The Maria The most prominent features of the Moon's surface were called maria, the Latin word for "seas," by Galileo because when he first looked at them through a telescope he thought they were seas. The maria consist of basalt flows that flooded the floors of huge impact basins. Unevenly distributed over 16 percent of the Moon's surface, the maria are concentrated on the Earth-facing hemisphere, and have only a fraction of the craters that have accumulated in the highlands.

Impact Basins Though associated with the maria, impact basins are separate features. The largest is the South Pole Aitken basin, 1,550 miles (2,500 km) wide, located on the far side. The Imbrium basin, about 1,118 miles (1,800 km) wide, and the Crisium basin, 684 miles (1,100 km) wide, are both found on the near side. The Orientale basin, 808 miles (1,300 km) wide, is located on the western limb.

The Regolith The incessant bombardment of lunar rocks by very small meteorites, called *micrometeorites,* creates a fine-grained soil called the *regolith.* The thickness of the regolith in the maria is 6.6 feet (2 m) to 26 feet (8 m) and may exceed 49 feet (15 m) in the highlands.

The dark plains regions called maria (below left) dominate the Moon's surface. The close-up of an astronaut's footprint in the lunar soil (right), photographed during the *Apollo 11* landing, indicates the consistency of the lunar soil.

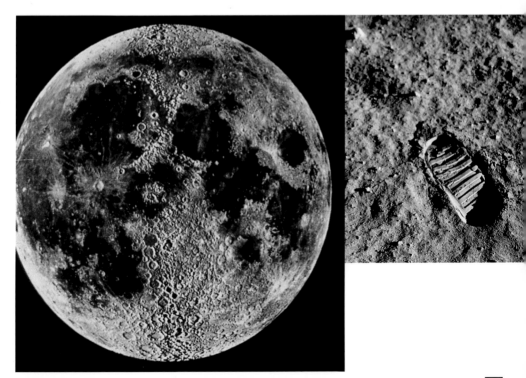

The Highlands The terrae, more commonly called the highlands or uplands, constitute about 84 percent of the lunar surface. Some of the highlands are mountains that form the rims of large basins—which were uplifted by the force of impact. The Apennine Mountains, part of the Imbrium basin rim, are among the highest mountains on the Moon. Much of the lunar highlands are elevated regions with dense concentrations of craters that occur in chains or clusters.

Volcanism

Lunar volcanism is confined to large impact basins, flooded by the mare basalts. The source of the mare basalts is over a hundred miles (160 km) inside the Moon's interior. Mare volcanism is confined to these basins because of the effect large impacts have on the crust. In a basin-forming impact, the crust is fractured to great depths and its thickness is reduced. Magma rises along the fractures beneath the basin and has less distance to travel to reach the surface. The basalts on the Moon can be most closely compared to flood basalts on Earth, although the lunar basalts have an average thickness that can be measured in hundreds of feet, reaching their greatest thickness near the basin centers and thinning toward the rims.

This photograph was taken on the *Apollo 15* mission of July 1971. Near the center of the Moon's Mare Imbrium, a lava flow is blocked by a wrinkle ridge. The spray of irregular craters in the center foreground was formed by material ejected from Copernicus Crater, which is 285 miles (460 km) to the south.

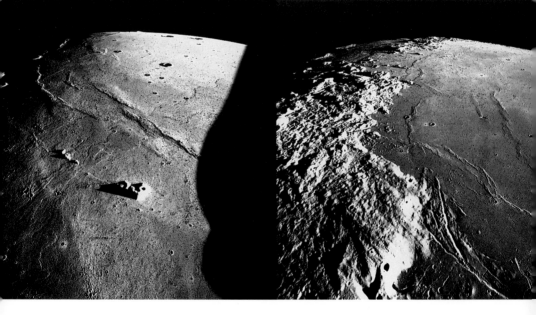

Apollo 17 captured this view of the west side of Mare Serenitatis in late 1972. Fractures are visible at the margin of the basin.

Tectonism

Lunar tectonism is concentrated in the maria, where the weight of the basalts causes the crust beneath to settle and subside. The basalts are stretched at the edges of the maria, causing fracturing and faulting that produce depressions called *graben*. The basalts on the basin interior are compressed, resulting in folding that produces wrinkle ridges analogous to the structures formed in the Caloris Basin on Mercury. Tectonic features in the lunar highlands are limited to small-scale scarps that appear to reflect small amounts of crustal shortening. They are similar to fault scarps on Mercury but much smaller.

Interior

The seismic measurements made at the *Apollo* landing sites provide clues to the interior structure of the Moon. The lunar crust varies in thickness from miles (tens of kilometers)beneath the maria on the near side to over 100 miles (160 km) in the highlands on the far side. The global average is about 44 miles (70 km). The upper crust has been broken, or *brecciated,* to depths that may be greater than 10 miles by impacts.

The Moon's mantle probably makes up much of its interior. The size, nature, and even existence of a lunar core have not been established. Indirect evidence suggests that if an iron-rich core does exist, it could be no larger than about 560 miles (900 km) in diameter and might constitute about 4 percent of the total mass of the Moon.

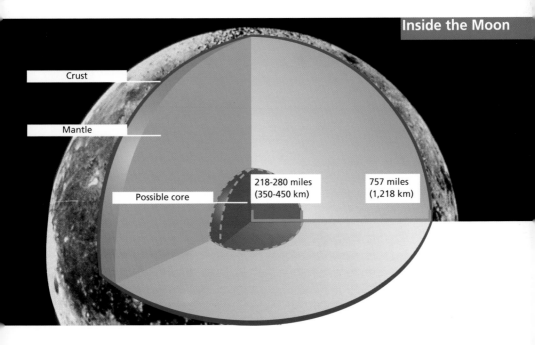

Inside the Moon

Crust

Mantle

Possible core

218-280 miles
(350-450 km)

757 miles
(1,218 km)

Phases of the Moon

The Moon, Earth's only natural satellite, has always been a source of fascination. As people have observed since ancient times, the Moon always shows the same face to Earth, but the portion of that face that is illuminated by the Sun changes. Thus the shape of the illuminated portion as seen by an observer on Earth also changes. Ancient cultures explained these events with myths and wor-

shiped the Moon ever endars accurately pr phases.

The Moon takes e (27.3 days) to rotate o orbit Earth. Thus the s us, and we always see tures. In its orbit arou Moon's surface is alwa Sun. The different ph

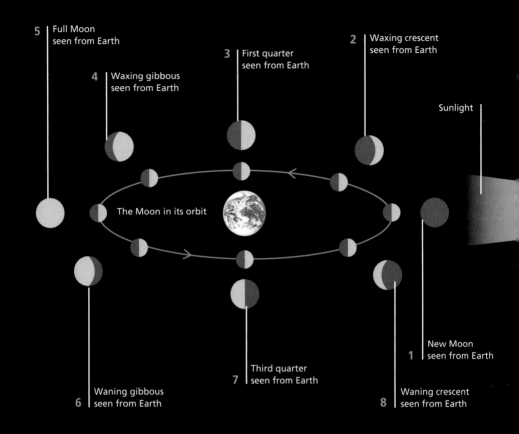

5 | Full Moon
seen from Earth

4 | Waxing gibbous
seen from Earth

3 | First quarter
seen from Earth

2 | Waxing crescent
seen from Earth

Sunlight |

The Moon in its orbit

1 | New Moon
seen from Earth

6 | Waning gibbous
seen from Earth

7 | Third quarter
seen from Earth

8 | Waning crescent
seen from Earth

viewed by an observer from Earth are numbered 1 through 8 on the diagram above. These phases are created by the changing angle between the Sun, Moon, and Earth, which varies in a regular cycle. The darkness of a new Moon occurs when none of the lunar surface can be seen from Earth. This is followed by a waxing (growing) crescent, then by the first-quarter phase, which is visible when the Moon has completed one-fourth of its orbit. Next is the waxing gibbous phase, then a full Moon when half of its orbital path around Earth is completed. After the full Moon, a gradual dwindling in its visible surface causes it to progress from the waning (decreasing) gibbous phase to a half-Moon at third quarter to the waning crescent. Finally, we see the darkness of a new Moon again before the cycle is repeated.

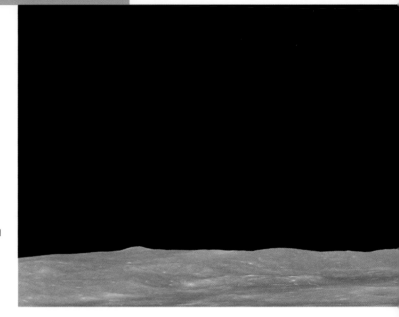

This spectacular view of Earth rising on the Moon's horizon was taken from the *Apollo 11* spacecraft in July 1969. The lunar terrain pictured is on the near side in the area of Smyth's Sea. A 1955 advertisement (below) offered land on the Moon for $1.00 per acre (2.47 ha).

The Origin of the Moon

The formation of the Moon may have been a unique process, unlike any other in the solar system. This is suggested in part by the large size of the Moon relative to Earth, leading some to describe Earth and Moon as a double planet.

Three competing hypotheses were popular before the Apollo missions cast doubt on all of them. The *co-accretion hypothesis* proposed that Earth and the Moon formed together as a double planet from the same gas and dust in the same location of the solar nebula. The *fusion hypothesis* held that Earth at one time began to spin so fast that a large piece broke away and formed the Moon. And the *capture hypothesis* proposed that the Moon formed elsewhere in the solar system and was captured by Earth's gravity when it traveled too close to Earth—either intact or as fragments torn apart by gravitational forces.

Analysis of lunar rocks has provided important insights into the Moon's origin. We now know that the average composition of the Moon and the composition of Earth's upper mantle are similar. The Moon and Earth have identical ratios of certain oxygen isotopes, and this indicates that they formed in the same part of the solar nebula.

But the differences are also significant. The Moon is relatively poor in volatile elements (elements that melt at low temperatures) and metallic elements such as nickel and cobalt, that are found in iron-bearing minerals. Also, lunar basalts are about ten

times richer in the *refractory* element (an element that remains solid at high temperatures) titanium than basalts on Earth.

If the capture theory were correct, the oxygen isotope ratios would probably be very different. If the fusion hypothesis were correct, the composition of the lunar rocks and Earth's crust and mantle should be more similar. And, if the co-accretion hypothesis were correct, the composition of the rocks would probably be more similar, as would the mean densities of the two bodies.

The latest popular model is the *giant-impact hypothesis*. More consistent with the available data than any of the other hypotheses, it proposes that Earth collided with a body possibly as big as Mars about 4.5 billion years ago. The mantles of both planets were almost totally vaporized by the impact, and the iron core of the smaller planet, which contained other metallic elements as well, settled through the lower mantle of Earth and became part of Earth's core. The debris—primarily material from the mantles of the two planets—blasted into space, and the heat of the collision vaporized the volatiles, including water. The remaining materials, rich in refractory elements and depleted of volatile elements, coalesced to form the Moon.

The giant-impact hypothesis may explain why the Moon's orbit is not in the plane of either Earth's orbit or its equator. The "impact" itself, presumed to have been delivered off center, may explain the Earth's spin.

Astronaut David R. Scott, commander of the *Apollo 15* mission, stands on the slope of Hadley Delta. The lunar roving vehicle is shown in the background, flanked by the Apennine Mountains.

Astronaut James B. Irwin operates the lunar roving vehicle during the *Apollo 15* landing. Mount Hadley is in the background. Rille Herigonius, shown at close range (above right), may be an ancient lava channel.

History

After the collision of a Mars-sized body with Earth, the accretion or gathering of material was so rapid that it generated an enormous amount of heat, almost completely melting the surface of the Moon to a depth of over 100 miles (several hundred km). The less dense calcium and aluminum silicates floated to the surface and formed the highlands. The denser iron and magnesium silicates sank and formed the mantle. Later, the mare basalts emerged from the mantle, flooding the lunar basins.

The magma ocean cooled slowly and had completely solidified by about 4.3 billion years ago. During the period of heavy bombardment, the highland rocks were pulverized, leaving craters and large impact basins all across the surface. At the end of the bombardment about 3.8 billion years ago, the Imbrium and Orientale basins formed. These were the last of the major impact basins. Around this time, magma rose along deep fractures in the lunar crust and flooded the floors of the impact basins, forming the maria. By about 3 billion years ago, the impact rate had slowed and became more or less constant. Flooding of the maria all but ceased, and the Moon became geologically inactive.

If you were there

If you weighed 100 pounds (45 kg) on Earth, you would weigh about 16 pounds (7 kg) on the Moon.

A Moon day would be two Earth weeks.

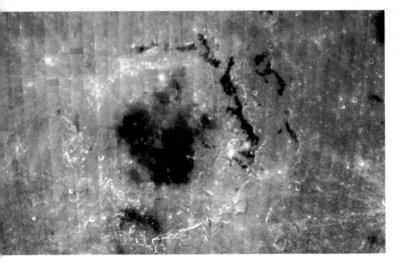

This digital mosaic of Mare Orientale, one of the Moon's large basins, was generated from images returned by *Clementine*. The outer ring, the Cordillera Mountains, is about 559 miles (900 km) wide.

Rotation and Revolution

Because the Moon rotates once during a single orbit around Earth, we always see the same side of the Moon from Earth. It also means that if you were standing on the Moon, Earth would appear to be motionless and fixed in the sky.

The Moon makes one revolution in relation to a fixed star in 27 days, 7 hours, and 43 minutes. This is called the *sidereal* month because it is measured with respect to the stars. The *synodic* month, which measures the period from one new moon to another, is 29 days, 12 hours, and 44 minutes. In the time it takes the Moon to journey around Earth once, Earth passes through one-twelfth of its orbit, or about 30 degrees. If the Moon starts out full, it takes one sidereal revolution plus about 2 more days before it appears full again.

The time of moonrise on Earth also depends on the motion of both the Moon and Earth. As Earth turns on its axis every 24 hours, the Moon moves 13.2 degrees in its orbit from west to east, making moonrise about 50 minutes later each day.

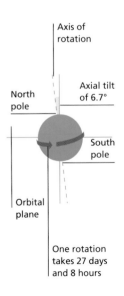

Axis of rotation

Axial tilt of 6.7°

North pole

South pole

Orbital plane

One rotation takes 27 days and 8 hours

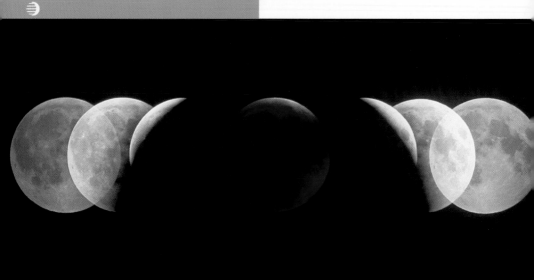

This composite of a total lunar eclipse shows the Moon moving from right to left into Earth's umbral shadow and then out again. The Moon is illuminated by the deep red color characteristic of many lunar eclipses.

Eclipses

The diameter of the Sun is about 400 times greater than the Moon's, but the Sun is also on average 400 times farther away from Earth. Hence the Sun and the Moon appear to be about the same size, which explains the phenomenon of solar eclipses. When the Moon comes directly between Earth and the Sun, its shadow is cast on Earth. At a certain distance from Earth, the Moon exactly covers the Sun's photosphere. During the *total* eclipse that follows, the Sun's corona flares, and prominences are visible.

When the Moon is farther from Earth, its apparent size is smaller and it may not completely cover the photosphere. This phenomenon is called an *annular* eclipse, in which a narrow rim of the Sun can still be seen.

If the Moon's orbit were exactly in the plane of Earth's orbit, called the *ecliptic,* there would be an eclipse of the Sun at each new moon and an eclipse of the Moon at each full moon. However, the Moon's orbit is inclined 5 degrees to the ecliptic, which is enough to make eclipses rare. There are two to five every year, each one visible only along the narrow path the Moon's shadow makes across Earth's surface. In any one locality, a total solar eclipse is visible only about once in 350 years.

Eclipses of the Moon by Earth's shadow are less frequent, but when they occur they are visible on over half Earth's surface. The maximum number of total lunar eclipses possible in any one year is three and the minimum is zero.

Tides

Tidal forces represent the differences in gravitational attraction exerted by an object like the Moon on parts of another object, like Earth. Although they are not strong enough to greatly influence the shape of Earth, tidal forces do influence the oceans by causing the tides. The more distant Sun also exerts a tidal attraction, although a weaker one. The time between high tides is always about 12 hours and 25 minutes—half the time from one moonrise to the next.

The extent of the tides varies, depending in part upon whether the Moon is nearer or farther from Earth. When the Sun and the Moon are aligned, their combined influence creates the highest and lowest tides, called *spring tides*. The *neap tides* occur when the Sun and the Moon are at right angles and exert the least influence.

The tidal forces exerted on the Moon by Earth are much greater. The rising and falling of "solid tides" has led to *tidal coupling* of the Moon to Earth and permanent solid bulges on the Moon. This tidal coupling is why the same side of the Moon always faces Earth.

Spring Tides

Neap Tides

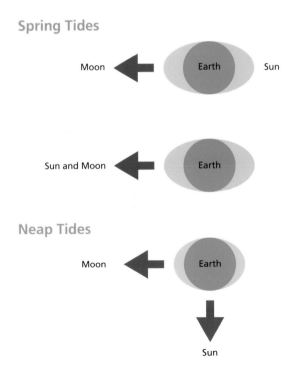

Tidal Forces

The gravitational attraction between two bodies in space creates tidal forces. The ocean's tides are caused by the tidal force between Earth and the Moon, but a body need not be covered with water to show the effects of this phenomenon. A tidal bulge is the distortion of a body's spherical shape owing to the gravitational pull of a nearby object. All planets exert tidal forces on their satellites. Jupiter's fiery moon Io owes its constant volcanic activity to tidal forces, as well. As it orbits, Io is pulled in different directions by Jupiter and by another satellite, Europa. The shifting tidal bulges created by these forces cause Io's molecules to grind against one another, generating the tremendous heat that fuels its many volcanoes.

This U.S. Geological Survey global mosaic of Mars is generated by using images and color data returned by the *Viking* orbiters. The planet's south polar cap is visible in this projection.

Mars

More than any other planet in the solar system, Mars has sparked and sustained our curiosity. Early observers, drawn by its reddish color, named it after the Roman god of war. Toward the end of the eighteenth century, astronomers began to wonder whether intelligent life—or life in any form—might exist on Mars. With the age of planetary exploration, we found that the fourth planet from the Sun is largely cratered and arid. But we also found a world filled with features we had not anticipated. We are still asking whether Mars may have supported some form of life in the past and whether life that is familiar or alien to us, complex or primitive, may thrive there now.

Early Investigation

Tycho Brahe's observations of Mars provided Johannes Kepler with the foundation for his three laws of planetary motion. Galileo may have been the first astronomer to view Mars with a telescope. In 1659, the Dutch astronomer Christiaan Huygens drew features of Mars and estimated the length of its day, or its rotation period, to be about 24 hours. By 1666 Giovanni Cassini had discovered that Mars has polar caps.

After measuring the tilt of the axis in 1777, William Herschel deduced that Mars must have seasons like Earth's. At 24 hours, 29 minutes, and 22 seconds, he came within 13 seconds of the currently accepted figure for its rotation period. Observing the seasonal retreat and advance of the polar caps, Herschel concluded that they must be thin. He also determined that the existence of clouds in the planet's atmosphere explained changes in brightness. Speculation about life on Mars was fueled by his theory that dark areas on the planet were seas.

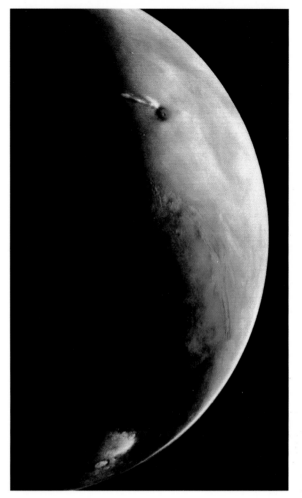

This image was returned by *Viking 2* as the spacecraft approached the dawn side of Mars in early August, 1976. The giant volcano, Ascraeus Mons, with a water ice cloud on its western flank, is on the left.

Schiaparelli In 1878, Giovanni Schiaparelli, a highly respected astronomer and director of the Milan Observatory in Italy, reported seeing a system of canals (*canali*) on Mars. He had made his observations during a *favorable opposition* of Mars in 1877. An opposition of Mars occurs when the orbits of the Sun, Earth, and Mars are aligned; a favorable alignment occurs about once every 17 years when Mars is near its perihelion, or nearest position to the Sun, and is only about 35 million miles (56 million km) from Earth.

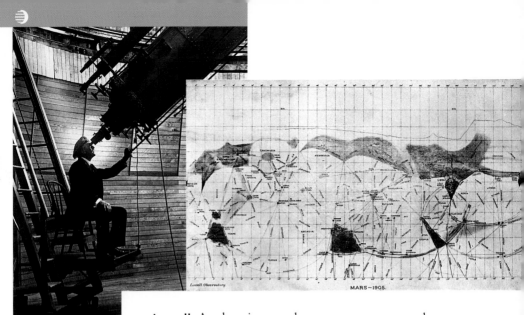

Astronomer Percival Lowell used a 24-inch Clark telescope (left) to observe Venus. His map of the planet Mars (right) was produced in 1905.

Lowell As the nineteenth century came to a close, a young mathematician, Percival Lowell, became intrigued by the popular concept of canals on Mars. Lowell built an observatory near Flagstaff, Arizona, at an altitude of 7,000 feet (2,128 m) to escape the distortion caused by Earth's atmosphere. He mapped five times more canals than Schiaparelli had reported. From this he concluded that some form of intelligent life had built the network of canals to circulate water from the wet polar regions to equatorial deserts. Lowell's belief, which he devoted much effort to publicizing, greatly increased science fiction writing centered on Mars, but left most astronomers of the time unconvinced.

Unpiloted Space Probes

Mars and Mariner The first spacecraft directed at Mars was launched by the Soviet Union. *Mars 1* flew by the planet Mars in November 1962 but failed to return useful data. Three years later *Mariner 4* was launched by the United States. It swept within 6,117 miles (9,842 km) of Mars and sent back 22 black-and-white images of a barren, cratered surface. Hopes for a diverse landscape on Mars were dimmed. Data indicated a weak magnetic field and an atmosphere so thin that it equaled about 1 percent of the atmospheric pressure at sea level on Earth.

In July 1969, *Mariner 6* flew within 2,120 miles (3,411 km) of the surface of Mars while the *Apollo 11* astronauts were still in quarantine on Earth after their trip to the Moon. *Mariner 7* reached Mars in August 1969. Together, the two probes returned a modest 199 images, which contained no new revelations. So in 1971 the next two Mars probes were readied, one to image the entire surface and the other to provide repeat coverage looking for any changes that might have occurred.

Mariner 8 plunged into the Atlantic Ocean shortly after blast-off, doomed by a second stage rocket failure, but in November 1971 *Mariner 9* became the first spacecraft to orbit another planet. Mars did not cooperate, however. A planet-wide dust storm that obscured most of the surface took weeks to clear. But when the dust settled, the summits of giant volcanoes appeared. The 7,000 images returned by *Mariner 9* finally revealed a landscape far more diverse than we had imagined.

One of four separate probes launched by the Soviet Union in 1973, *Mars 5* was among the most successful. It returned 70 images of the same quality as those returned by *Mariner 9*.

The deep rift valleys of the Valles Marineris canyon system scar the face of Mars. The giant canyon extends more than a fifth of the way around the planet.

Viking Missions

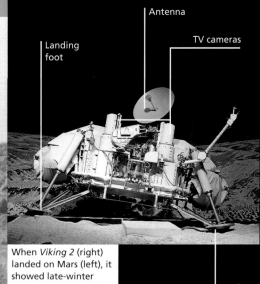

Antenna

Landing foot

TV cameras

Landing foot

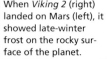

When *Viking 2* (right) landed on Mars (left), it showed late-winter frost on the rocky surface of the planet.

Each of the two unpiloted Viking spacecraft launched in 1975 consisted of an orbiter carrying a lander. They were the first spacecraft to conduct prolonged research on the surface of another planet.

As soon as *Viking 1* entered orbit around Mars in June 1976, it began sending images of potential landing sites to mission controllers on Earth. Guided by its onboard computers, *Viking 1* made the first successful soft landing of a spacecraft on Mars on July 20, 1976. Within minutes its cameras were activated, and long-awaited images of the planet's surface were received on Earth. These showed a desolate, rust-colored landscape of rocks and boulders with a reddish sky overhead. *Viking 2*, which landed on the opposite side of the planet, returned a similar view.

Operating until mid-1980, Viking orbiters imaged almost 100 percent of the surface of Mars, transmitted more than 55,000 images, and returned extensive data on the planet's atmosphere, temperature, and weather. The landers operated until 1983, transmitting more than 4,500 images and performing chemical analysis of the soil samples they collected. These analyses could not confirm the presence of organic matter or the existence of past or present life on Mars.

Viking Without question, the Viking probes of the 1970s, sent to detect life forms Mars might support, were the most ambitious and successful to date. Two spacecraft, each consisting of an orbiter and a lander, were launched by the United States in 1975. *Viking 1* reached Mars on June 19, 1976, and spent three weeks searching for a safe, appropriate landing site. On July 20, 1976, the *Viking 1* lander touched down on the plains of Chryse Planitia. Our first glimpse of the Martian landscape from the surface revealed a rocky, desertlike terrain. Two weeks later *Viking 2* reached Mars, and the second Viking lander set down in September 1976. Although it landed on the opposite side of Mars, the images it returned to Earth showed the same rocky surface revealed by *Viking 1*.

The landers recorded the physical properties and chemistry of the soil, monitored the landing site through the seasons, and gathered meteorological data. The orbiters imaged the entire surface, returning more than 55,000 images, some very high resolution. Both the orbiters and landers functioned flawlessly and outlasted their expected worklife. The *Viking 1* lander returned images and meteorological data until contact was lost in 1983. No conclusive evidence of life was found.

Mars' Utopia Planitia was the touchdown site for *Viking 2* lander in September, 1976. The terrain proved to be flatter, but more rock-strewn, than the Chryse Planitia region where *Viking 1* had landed a few weeks earlier.

The rocky surface of Chryse Planitia as viewed by the *Viking 1* lander. The prominent boulder in the foreground, named Big Joe, is over 3 feet (1 m) high.

Later Missions More recent missions have been unsuccessful. In 1988 the Soviets launched *Phobos 1* and *2* to study Phobos and Deimos, the moons of Mars. Contact with *Phobos 1* was lost early, but *Phobos 2* came within a few miles of success. Having reached Mars and gone into orbit, it was 76 miles (120 km) from Phobos. As the spacecraft attempted to move to within 22 miles (35 km) of that moon, contact with the probe was lost.

In 1992 the United States launched the *Mars Observer* in an attempt to provide the most detailed topography and the highest resolution images that had ever been obtained. Contact was lost, however, just as the spacecraft was preparing to enter orbit around Mars in August 1993.

General Characteristics

A little over half the diameter of Earth, Mars resembles Earth more than any other planet does. Its day is 24 hours 37 minutes long. Its axis is tilted 25.2°, nearly the same as Earth's. In common with Earth, Mars has seasons. But, despite its many similarities, Mars is still a very different world.

Orbit, Temperature, and Gravity

The orbit of Mars is the third most elliptical of the planets. Mars is about one and a half times farther from the Sun than Earth. At its perihelion, Mars is 129 million miles (208 million km) from the Sun. At its aphelion, it is 155 million miles (249 million km) from the Sun. The difference between the two is about 26 million miles (46 million km), whereas on Earth the difference between perihelion and aphelion is only about 3 million miles (5 million km). Because of its greater average distance from the Sun and the wider range of its nearest and farthest approach to the Sun, Mars has a lower average surface temperature than Earth's. It also has greater variations in temperature.

At perihelion, Mars receives about 45 percent more solar radiation than at aphelion. In winter, temperatures at the planet's south pole fall to about 193°F below zero (−125°C), the freezing point of carbon dioxide. And during summer in the southern mid-latitudes, noontime temperatures reach 72°F (22°C).

Even though Mars is more than half the diameter of Earth, it has only about a tenth of its mass. The mean density of Mars is the lowest of the terrestrial planets—3.9 as compared to Earth's 5.5. This suggests that its iron-bearing core is proportionately smaller than Earth's. Its gravitational field is a little over a third that of Earth's, so a person on Mars would only weigh about one-third of their weight on Earth.

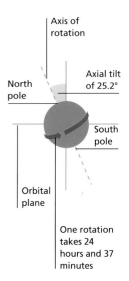

Axis of rotation

Axial tilt of 25.2°

North pole

South pole

Orbital plane

One rotation takes 24 hours and 37 minutes

The Martian Atmosphere

The composition and density of the Martian atmosphere are very different from Earth's, but there are many similarities. Like Earth, Mars has clouds, weather patterns, and a predominant wind direction. But its atmosphere is made up of about 95.3 percent carbon dioxide, 2.7 percent nitrogen, 1.6 percent argon, and less than 0.2 percent oxygen. It contains about 0.03 percent water vapor—close to the limit it can hold. The hazes and fogs often seen in low-lying areas at dusk and dawn are evidence that the atmosphere is near its saturation point with water vapor.

Even though the Martian atmosphere, which is now less than 1/100th the density of Earth's, may have been much denser at an earlier time, the relatively weak gravitational field would probably not have held onto it for very long. Data returned by the *Phobos* spacecraft suggest that the solar wind is carrying away the weakly held atmosphere at a rate of 50,000 tons (45,400 MT) a year. If this is true, and if the planet had only a weak magnetic field to shield it, much of the early, denser atmosphere might have been quickly swept away.

The recession, or shrinking, of the south polar cap of Mars, shown at the top of these images, is evident between the Martian spring (left) and Martian summer (right).

As much as a third of the atmosphere may be frozen at the poles at times. Evaporation, condensation, and freezing of carbon dioxide and water vapor create changes in the sizes of the polar caps.

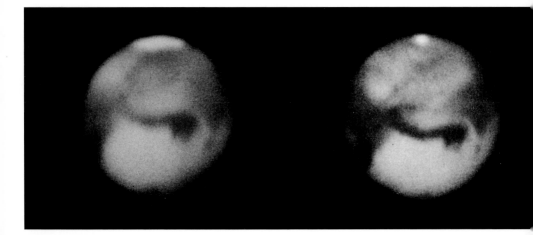

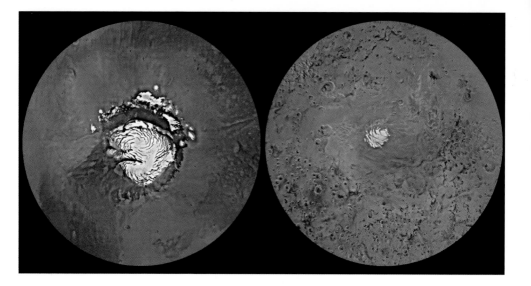

The Surface of Mars

The *Viking* landers provided spectacular views of the surface of Mars. The two lander sites, on different sides of the planet, are remarkably similar. Both are covered with rocks ranging from pebbles to boulders.

The *Viking 1* lander site, in Chryse Planitia near the Martian equator, has gently rolling terrain. Fine-grain soil has been blown by the wind into dunes, creating a landscape not very different from places in the Western Desert of Egypt. The *Viking 2* lander site in Utopia Planitia, in the northern lowlands, is generally flat and featureless out to the horizon. Although fine-grain material is present between the rocks, dunes have not formed here.

The rocks at the two sites may be volcanic in origin and basaltlike in composition, and they may have been excavated by impacts. An iron-rich clay, similar to the clay produced by the weathering of basalt, is the primary component of the soil, giving the soil its orange-red color.

These are digital mosaic images of the north (left) and south polar caps of Mars. Even though the south polar cap is smaller, it remains throughout the Martian summer.

If you were there

If you weighed 100 pounds (45 kg) on Earth, you would weigh about 38 pounds (17 kg) on Mars.

You could not breathe the air and would need protection from the cold for most of a Mars year.

Daytime temperatures may reach 72°F (22°C) in some places, but your skin would burn severely without a protective ozone layer.

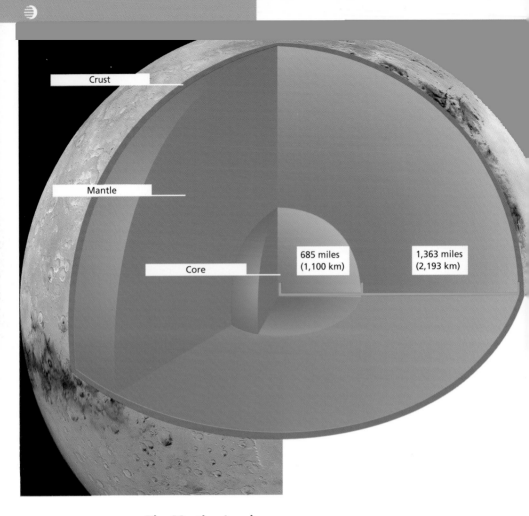

Crust

Mantle

Core

685 miles
(1,100 km)

1,363 miles
(2,193 km)

The Martian Landscape

Like the other terrestrial planets, Mars has highlands and low-lands. The distribution of the terrain, however, is unique: the highlands occur in the southern hemisphere and the lowlands in the northern hemisphere. The highlands are composed of the old-est, most heavily cratered crustal material on the planet. They rise .6 to 1.9 miles (1 to 3 km) above the Martian *topographic datum*, a level analogous to sea level on Earth. They were sub-jected to the same heavy bombardment that is recorded on the surfaces of the Moon and Mercury.

The oldest highland terrain—older than the end of the period of bombardment 3.8 billion years ago—is etched with a network of small channels that may have been carved by flowing water from rainstorms. Flood volcanism occurring at about the same time formed the smooth plains that lie within the heavily cratered highlands. The smooth, generally featureless plains dominating

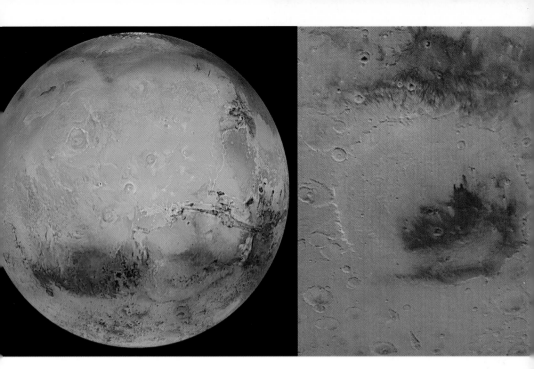

the northern lowlands lie below Martian "sea level" and have relatively few craters. They formed after the period of bombardment, when the rate of cratering became lower. The boundary between the lowlands and highlands is marked by an escarpment, or long cliff, and in some areas, irregularly shaped mesas—isolated remnants of the highlands—rise above the lowland plains. It has been suggested that the great differences in the highlands and lowlands may have been the result of a giant impact.

Tharsis Rise A prominent exception to the general division between the highlands and the lowlands is the Tharsis Rise. About 6 miles (10 km) high and 5,000 miles (8,000 km) across, it forms a distinct bulge that dominates the planet's western hemisphere. The Tharsis Rise has the greatest concentration of volcanic and tectonic features on Mars. Huge shield volcanoes, extensive fracture and ridge systems, and the enormous Valles Marineris canyon system all appear to be linked to the rise.

Impact Basins Mars, like the Moon and Mercury, has large impact basins. The largest, Hellas, with a diameter of about 1,200 miles (2,000 km), may be the biggest impact basin in the solar system. The Isidis Basin, located on the boundary between the highlands and the lowlands, is about 1,180 miles (1,900 km) across and appears to have been flooded by the same volcanic material that covers the lowlands. The best preserved of the largest basins, Argyre, is about 750 miles (1,200 km) wide.

The global mosaic of Mars (left) is centered on the Tharsis Rise. This region has been the core of much of the volcanic and tectonic activity on the planet. Mars's enormous impact basin, Hellas (right), dominates the landscape.

The Interior Structure

Because of a lack of seismic data for Mars, we know little of the interior structure. Estimates of the core size vary, some going as high as 1,250 miles (2,000 km). The predominant theory is that the iron-bearing core has a radius of about 680 miles (1,100 km), which makes up about 6 percent of the planet's mass. In contrast, Earth's core accounts for about 32 percent of its total mass. The weak magnetic field suggests that the core is no longer liquid or that currents within a liquid core are slow.

The mantle is estimated to extend no deeper than about 1,400 miles (2,300 km) and to have an average density of 4, as compared to 6 for the core. In the Tharsis Rise, the thickness of the crust and the lithosphere varies from about 12 miles (20 km) to 90 miles (150 km). This may explain at least in part why plate tectonics did not develop on Mars.

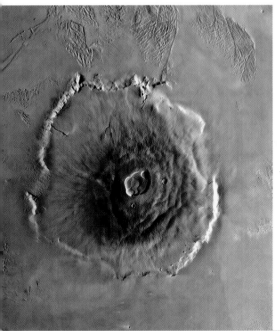

Olympus Mons

This digital mosaic shows Olympus Mons, the largest volcano on Mars that towers 15 miles (24 km) above the surrounding plains and 16 miles (26 km) above Martian topographic datum. The central caldera is about 55 miles (90 km) across. The steep cliffs at the base of the volcano are 3.5 miles (6 km) high.

This outline of the continental United States was superimposed on a U.S. Geological Survey shaded relief map of the Tharsis Rise. Drawn to scale, it illustrates the size of the Tharsis volcanoes.

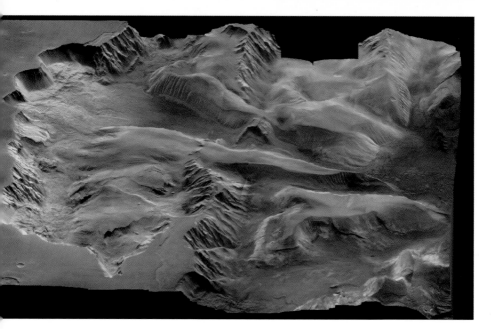

Volcanism

Mars may be one of the smaller terrestrial planets, but it has some of the largest volcanic and tectonic features in the solar system. Many of these features are found on the Tharsis Rise. The huge bulge that forms the rise may be the result of a broad uplift of the crust caused by upwelling of mantle material or of buildup of the crust from many episodes of flood volcanism, or both.

The shield volcanoes in the Tharsis Rise are the largest in the solar system, and the biggest of them all is Olympus Mons—named after Mount Olympus, the highest mountain in Greece and home to the gods of Greek mythology. Some 375 miles (600 km) across and more than 16 miles (26 km) high, Olympus Mons would cover the state of Washington and nearly half of Oregon. It has a prominent summit caldera complex and a cliff that is 3.5 miles (6 km) high at its base. The amount of lava erupted to form Olympus Mons, and its rate of flow may have been from 10 to 100 times greater than that of any volcano on Earth.

Three other large shield volcanoes—Arsia Mons, Pavonis Mons, and Ascraeus Mons—are each over 12 miles (20 km) high and form part of a volcanic chain near the center of the Tharsis Rise. Along the rise's northern edge sits Alba Patera, over a mile high and about 400 miles (700 km) across. Formed by both volcanism and tectonism, Alba is surrounded by a well-developed fracture system and is the source of lava flows extending up to 1,000 miles (1,600 km) from its center. It may be an unusual low-relief volcano or a feature similar to coronae on Venus. Smaller volcanoes are found in the highlands and in the Elysium province in the northern lowlands.

Perspective view of Ophir (left) and Candor (right) Chasma, part of the Valles Marineris canyon system.

Future U.S. Missions to Mars

An artist's conception sequences the descent of the Mars Pathfinder spacecraft through the Martian atmosphere. At upper left, the craft decouples from a cruise stage and heats up as it enters the atmosphere. Then a parachute is deployed to slow the craft.

After the parachute has deployed and the heat shield released, Pathfinder is lowered on a tether that extends from 66 to 132 feet (20 to 40 m). Then the lander, lower right, opens its petals and retracts the air bags in preparation for surface operations.

From 1996 to 2005, the United States will engage in the Mars Surveyor program. This large program plans to launch spacecraft to Mars every 26 months, when Mars is aligned with Earth. The Mars Pathfinder and the Mars Global Surveyor missions are part of the Mars Surveyor program.

Mars Pathfinder is scheduled for launch in December 1996. It will arrive on Mars in July 1997, at a carefully selected site. The site will probably be located on an ancient flood plain not far from the *Viking 1* lander that touched down on Mars in 1976. This region has a low elevation.

Pathfinder will be powered by sunlight collected by its solar panels. During descent, it will use the aerobraking technique to slow its speed. Aerobraking involves using the friction generated by passing through a planet's atmosphere to slow a spacecraft's speed. The success of this technique will be carefully analyzed for use in future missions.

In this rendition, the lander's open petals are covered with solar cells to provide electrical power for the surface operations. The gold thermal blanket protects lander electronics from the temperature extremes on Mars. A mounted camera is in operation.

This artist's conception shows the landing of the first human mission to Mars in 2019. Astronauts carry a hand-held camera, as a dust storm approaches the cratered area near the landing site. Mars's moons, Phobos and Deimos are visible in the twilight sky.

When Pathfinder descends through the Martian atmosphere, a parachute will open to slow it down before landing. The lower elevation of the landing site will allow more time for the parachute to work. The lander's camera will provide a panoramic view of the landing site. Once on Mars, a sophisticated rover vehicle deployed from the lander will scour the landing site analyzing the rocks and soil. The Pathfinder lander will monitor weather on the planet.

The Mars Global Surveyor is scheduled to leave Cape Canaveral in November 1996 and arrive at Mars ten months later. Once in orbit, Global Surveyor will begin mapping the surface of Mars in January 1998. It will return data allowing the first detailed topographic maps of Mars to be made, analyze the distribution of minerals, and determine Martian weather patterns. Global Surveyor will continue mapping for about two years, after which it will serve as a relay station for other landers sent to Mars over the next three years.

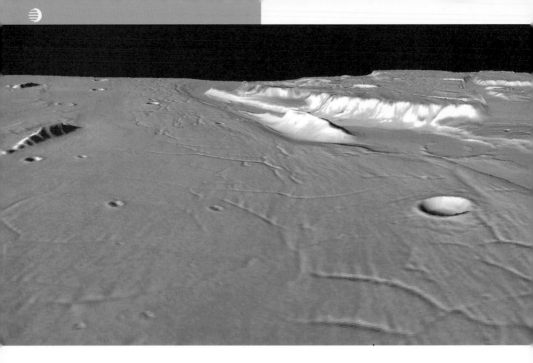

This three-dimensional perspective view of Kasei Valles was compiled from *Viking* orbiter images and topographic data.

Hawaii's Mauna Loa, the largest volcano on Earth, is 5.7 miles (9.1 km) above the ocean floor and is 75 miles (120 km) across. Why are the Martian volcanoes so much larger? The answer may be that Mars, unlike Earth, is a one-plate planet. If this is so, heat is lost from the interior by conduction through the lithosphere and crust and by advection through hot-spot volcanism. On Earth, a stationary hot spot beneath the moving Pacific plate supplies the magma that formed Mauna Loa and the other Hawaiian volcanoes. With time the plate will move over the hot spot and eventually a new volcano will form some distance away in the same chain. If Mars is a one-plate planet, the plate never moves, and the magma supplied by the hot spot allows the volcano to grow until the hot spot itself disappears.

Other indications of hot spot volcanism on Mars are the extensive volcanic plains that appear to be formed by flood volcanism. Some of the largest plains, in the Tharsis Rise, were probably formed about 3.5 billion years ago and are many times greater than the largest flood provinces on Earth. Although their composition is uncertain, the volcanic plains most likely consist of basalt or basaltlike flows.

Tectonism

The tectonic features on Mars are no less impressive. In the probable absence of plate tectonics, these features appear to be distributed over broad regions rather than along plate boundaries as on Earth. The volcanic plains and highlands within the

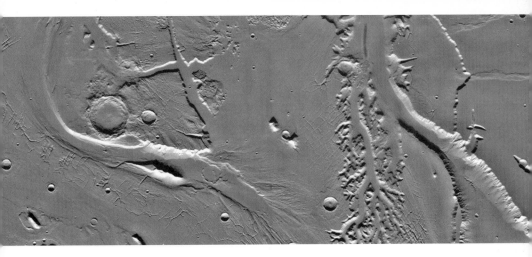

Tharsis Rise have been heavily fractured and faulted. The systems of fractures and graben form radial patterns that converge on an area very close to the center of the Tharsis Rise, much like the spokes of a bicycle tire. The great Valles Marineris canyon system falls along such a spoke.

This is a view from a spacecraft showing part of Mars's huge channel, Kasei Valles. The channel cuts deep into the volcanic plains and extends over a thousand miles (2,000 km) into Chryse Planitia. Note the wrinkle ridges in the volcanic plains.

Valles Marineris Named after the *Mariner 9* probe that discovered it, the stupendous Valles Marineris canyon system is up to 5 miles (8 km) deep and 2,800 miles (4,500 km) long—as long as the continental United States and large enough to hold the Alps or the Rocky Mountains. Its smaller tributary canyons are as large as the Grand Canyon in the United States. Valles Marineris may have been formed by the rifting, or the pulling apart, of the Martian crust, possibly resulting from the formation of the Tharsis Rise. It may be analogous to the East African and other continental rifts on Earth, but on a much larger scale.

Other tectonic features on Mars reflect shortening of the crust. The volcanic plains surrounding the Tharsis Rise have been buckled into a series of wrinkle ridges, like those formed on the other terrestrial planets and the Moon. Unlike the fractured radial patterns formed on the rise, these ridges form a circular pattern, much like the tire of a bicycle wheel. The most common tectonic feature on Mars, as on Venus, wrinkle ridges occur in volcanic plains outside the Tharsis Rise as well.

Within the highlands of Mars, scarps, or cliff-like features similar to those found on Mercury and the Moon, also reflect crustal shortening. One of the largest of the Martian scarps, Amenthes Rupes, is over 250 miles (400 km) long and up to 2 miles (3 km) high, comparable in size to the Discovery Rupes on Mercury. A scarp this big may indicate over a mile of crustal shortening. Like scarps on Mercury, those on Mars may reflect some degree of global contraction of the crust due to cooling of the interior.

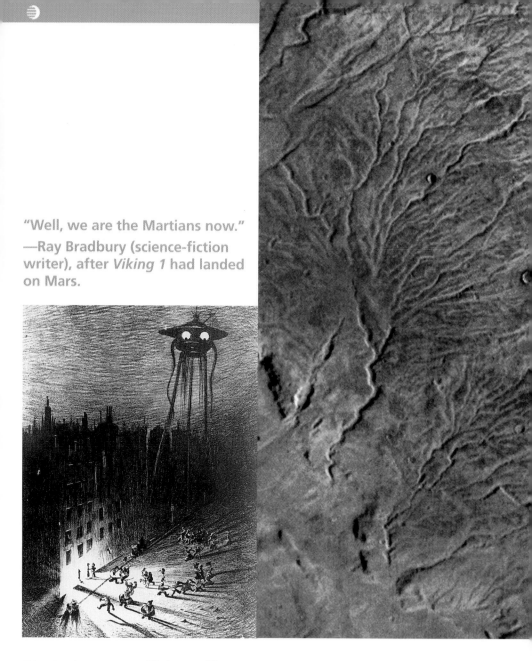

"Well, we are the Martians now."
—Ray Bradbury (science-fiction writer), after *Viking 1* had landed on Mars.

This artist's depiction of a Martian invasion of London is an illustration in the classic novel, *The War Between the Worlds*, by H. G. Wells.

Water on Mars

Many landforms on Mars seem to have been created or modified by flowing water. The channels that form valley networks in cratered highlands are reminiscent of streams on Earth. Much larger channels like Kasei Valles also exist. Kasei Valles cuts over a mile deep into the volcanic plains of the Tharsis Rise and extends some 1,250 miles (2,000 km) into Chryse Planitia, the *Viking 1* lander site.

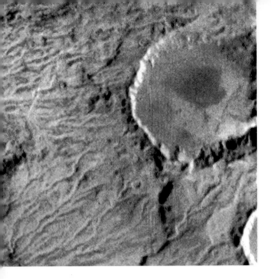

Viking orbiter returned this image of valley networks in the Martian highlands. Because the dendritic (dry river bed) channels are characteristic of drainage networks, rather than major flooding, they are thought to have been formed by drainage of precipitation.

Other large, irregular channels, emerging from what geologists call "chaotic" areas, intersect with the eastern end of the Valles Marineris. These channels extend for thousands of miles into Chryse Planitia. Features such as scour marks and teardrop-shaped landforms in Chryse Planitia suggest that huge volumes of water once flooded and extensively eroded the area.

Some features like the valley networks in the highlands may have been formed by rain when the atmosphere was denser and the climate was warmer. Or they may have formed as water trapped underground beneath a permafrost layer was gradually released. The large channels that flow into the Chryse basin may have been carved by great floods, in the same way the Channeled Scablands in the eastern part of the state of Washington were formed. About 18,000 years ago, an estimated 480 cubic miles (2,000 cu km) of water were released from an ice dam, creating the catastrophic flood that formed the Channeled Scablands. If the Martian channels were formed by similar floods, they must have been 10 to 100 times greater.

If water formed the valley networks and channels on Mars, where is it now? Most of the water is believed to exist as permafrost in the northern lowlands, as permafrost and possibly groundwater in the heavily fractured and cratered highlands, and as ice at the poles.

Polar Caps

The polar regions of Mars are covered with layered deposits of water ice, carbon dioxide frost, and dust. The layered deposits have eroded, leaving deep valleys that spiral outward from the poles. Although smaller in the summer, the frozen carbon dioxide cap in the south remains. During summer in the north, carbon dioxide returns to the atmosphere, leaving a cap of water ice.

Life on Mars?

The *Viking 1* and 2 landers were designed to search for life—however primitive—on Mars. They analyzed soil samples that their mechanical arms scooped from the planet's surface, testing for organic compounds—containing carbon—that might indicate biological origin, as they do on Earth. Meteorites also contain organic compounds, so it was assumed that they would be found on Mars. Yet analysis of the soil chemistry revealed no organic compounds.

The *Viking* landers sampled soil in different sites and got similar results. Thus no conclusive evidence of life was found. It has been proposed that no organic compounds, even those introduced by meteorites, were found on Mars because ultraviolet light reacts with the soil to form peroxide. This in turn reacts with any carbon present to form carbon dioxide.

Perhaps life exists under the permafrost or under the ice caps. On Earth, some algae, bacteria, and fungi have been found in ice-covered lakes in Antarctica, and further exploration of Mars may reveal that its ice harbors similar life forms.

American astronomer Asaph Hall discovered Phobos (below) and Deimos (right) in 1877. In this *Viking* mosaic image, Stickney, the largest crater on Phobos, appears in the upper left. It was named after Angelina Stickney, Hall's wife.

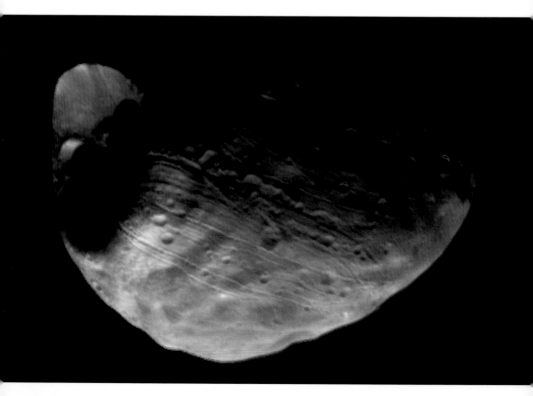

Phobos and Deimos

So small that they were not discovered until 1877, the satellites of Mars, Phobos and Deimos, have irregular shapes, low density, and a great many craters. They reflect so little light (low albedo) that they are among the darkest objects in the solar system. Like most satellites, they orbit in a west to east direction.

Potato-shaped Phobos measures about 13 miles (21 km) at its widest and has an average density of only 2.2. It orbits Mars in about 7 hours and 39 minutes, only 3,700 miles (6,000 km) away. It always has the same face turned to the planet, like Earth's Moon. If you were standing on Mars, you would see Phobos rising in the west twice each day.

Deimos measures about 7 miles (12 km) across and has an average density of only 1.7. Covered with a thick layer of dust, it orbits 12,500 miles (20,000 km) above the planet. Its orbit takes about 30 hours and 18 minutes, almost the same as a Martian day, so it appears to rise in the east and move across the sky for two and a half days before setting.

It is believed that these two planetary satellites are asteroids that were captured by Mars's gravitational field. Both are probably covered by a thick soil, like most asteroids. While both are heavily cratered, Phobos has one very large crater called Stickney that is 6 miles (10 km) wide. The grooves that line the surface may be related to that impact.

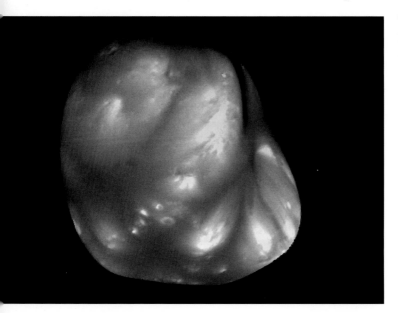

With the sunlight shining directly on the satellite Deimos, its irregular shape is clearly visible.

The Outer Planets

With advances in space technology, the outer planets have begun to reveal their secrets. We have watched comet fragments rain down upon mighty Jupiter, analyzed the spectacular rings of Saturn, discovered the peculiar sidelong spin of Uranus, and studied the stormy atmosphere of deep-blue Neptune. Even far-off Pluto, orbiting with its moon Charon on the outskirts of the solar system, is within our reach. And the planets are only part of the story—many moons are as complex and fascinating as the bodies they orbit.

Three days after flyby, *Voyager 2* captured this final, haunting look at Neptune and its largest moon, Triton.

Jupiter, its Great Red Spot and one of its four largest moons, Io, are visible in this *Voyager 1* image, taken in early 1979.

Jupiter

Viewing it from a distance, someone unfamiliar with our solar system might describe it as a star orbited by one planet and eight other minor objects. That "one planet" would be Jupiter—by far the largest object in the solar system after the Sun. At 88,865 miles (142,984 km) in diameter, compared to Earth's diameter of 7,926 miles (12,753 km), Jupiter is so large that more than 1,000 Earths could fit inside it. In fact, all the other planets could easily fit inside it. With about two and a half times the mass of the other planets combined, it is estimated that Jupiter contains about 71 percent of all the material in the solar system, excluding the Sun. Thus the strength of its gravitational pull is second only to the Sun's. In many ways Jupiter and its system of 16 satellites are almost like a solar system in themselves.

Earth

"The exploration of the solar system is an extremely strong affirmation of being alive, of being curious about our place in space."
—Carl Sagan

Exploring Jupiter

Jupiter is the largest planet, and it is about 800 million miles (1.3 billion km) from Earth, on average. Even with the best telescopes, it is difficult to observe in much detail. It was not until spacecraft traveled to the outer planets that the complexity of Jupiter's atmosphere, the diversity of its moons, and its faint rings were discovered.

Pioneer Missions *Pioneer 10*, launched in March 1972, was the first attempt to reach Jupiter with a space probe. The first spacecraft to go beyond Mars, *Pioneer 10* flew by Jupiter on December 3, 1973, returning 23 low-resolution images of the Jovian cloud system. A year later, *Pioneer 11* returned 17 images during its nearest approach to Jupiter. It then used the planet's gravitational force to adjust its course and proceed to Saturn. These two probes also relayed data on the temperature and pressure within Jupiter's atmosphere and several very low-resolution images of its moons.

The Voyagers The United States launched *Voyagers 1* and *2* in 1977. They flew by Jupiter in March and July of 1979 before proceeding to Saturn. *Voyager 1* then headed out of the solar system while *Voyager 2* went on to Uranus and Neptune. These highly successful probes returned about 30,000 images and a wealth of other data (see "Voyager Missions," p. 132).

Galileo In 1989 the *Galileo* probe was launched by the United States to rendezvous with Jupiter in 1995. In July 1994, when fragments of the comet Shoemaker-Levy 9 hit Jupiter, *Galileo* was still about 140 million miles (225 million km) away from the planet, but it was able to relay images of the collisions back to Earth (see "The Shoemaker-Levy Impacts," p. 143).

When *Galileo* arrives at Jupiter, it will orbit the planet and release a probe into Jupiter's atmosphere. Its descent slowed by a parachute, the probe will return information about the Jovian atmosphere until it succumbs to the enormous temperature and pressure it will encounter. *Galileo* will continue to orbit Jupiter, collecting and relaying images of the planet and its moons, as well as other data, for two years.

In January 1972, NASA technicians make final adjustments to the *Pioneer* vehicle. *Pioneer* was the first spacecraft designed for travel into the outer solar system and to operate effectively there—as far away from the Sun as 1.5 billion miles (2.4 billion km).

> "Our sense of novelty could not have been greater had we explored a different solar system."
> —A member of the Voyager imaging-science team describing the Jupiter discoveries.

Jupiter: Protector of Earth?

A fascinating computer program that simulates the growth of a solar system was developed at the Carnegie Institution in Washington, D.C. Each time it was run, rocky inner planets, gaseous outer planets, and a great many icy comets appeared. In scenarios lacking a Jupiter-size outer planet, the rocky planets comparable to Earth were peppered with comet impacts. But when a Jupiter-size giant did develop, its enormous gravitational influence kept most comets and asteroids from reaching the inner planets. On the basis of these simulations, it appears that Earth would be hit by asteroids and comets much more frequently without Jupiter to shield us.

Jupiter's Interior

Like the Sun, Jupiter consists mostly of hydrogen and helium. It also contains small amounts of methane, ammonia, and water vapor. Because of Jupiter's great mass, which is about 318 times greater than Earth's, its interior temperatures and pressures are extremely high. Jupiter's average density is lower, however—only 1.3 compared to Earth's 5.5.

Jupiter's rapid rotation (9 hours, 55 minutes) makes it flatten at the poles and bulge at the equator, forming a shape known as an *oblate spheroid*. Jupiter's equatorial radius is about 2,700 miles (4,300 km) greater than its polar radius. The size of the bulge suggests the presence of a liquid interior because fluids respond more readily to strong centrifugal forces that are generated by rotation.

Based on its shape and its gravitational field, Jupiter must have a dense core. This core is estimated to be 10 to 20 times the mass of Earth. Because Jupiter emits about twice as much heat as it

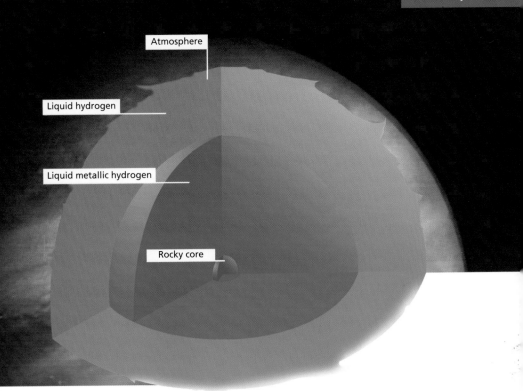

Atmosphere

Liquid hydrogen

Liquid metallic hydrogen

Rocky core

absorbs from the Sun, the core temperature is probably on the order of 36,000°F (20,000°C), about three times greater than the temperature of the Earth's core. This heat is thought to be generated as the planet slowly contracts by about an inch a year, compressing its interior. Jupiter's core is thought to be composed of melted rock or silicates and ice. The core pressure is estimated to be 100 million times the pressure on Earth's surface.

In a layer extending from 12,400 miles (20,000 km) below the cloud tops to the core, called the mantle, pressure causes hydrogen to take on the properties of a metallic liquid. In this state, hydrogen conducts electricity. The currents circulating in this layer are thought to generate the planet's powerful magnetic field, which is 20,000 times stronger than Earth's.

Surrounding the mantle of metallic liquid hydrogen is ordinary liquid hydrogen. Above this, in the outer portion of the interior, hydrogen behaves as a gas, circulating in large cells generated by the planet's rapid rotation.

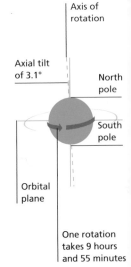

Axis of rotation

Axial tilt of 3.1°

North pole

South pole

Orbital plane

One rotation takes 9 hours and 55 minutes

Voyager Missions

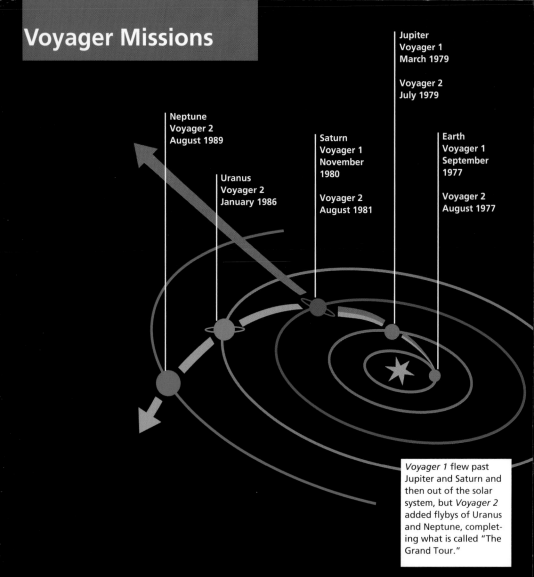

Jupiter
Voyager 1
March 1979

Voyager 2
July 1979

Neptune
Voyager 2
August 1989

Saturn
Voyager 1
November
1980

Voyager 2
August 1981

Earth
Voyager 1
September
1977

Voyager 2
August 1977

Uranus
Voyager 2
January 1986

Voyager 1 flew past Jupiter and Saturn and then out of the solar system, but *Voyager 2* added flybys of Uranus and Neptune, completing what is called "The Grand Tour."

On August 20, 1977, *Voyager 2* was launched from Cape Canaveral. Sixteen days later, *Voyager 1*, which would take a more direct route to Jupiter and Saturn, was launched.

Each Voyager was equipped with three programmable computers and instruments to conduct scientific investigations. High resolution cameras were mounted on a steerable platform.

Initial technical problems were solved as computer adjustments transmitted to the Voyagers corrected some erratic spacecraft behavior. Then, in March and July 1979, the Voyagers passed Jupiter in the first of a series of encounters that would last for more than a decade. *Voyager 1* traveled on to Saturn, first reaching it in November 1980. Its trajectory took it behind Saturn's moon Titan then out of the solar system by 1990.

Each Voyager is equipped with two television-type cameras (above). The artist's conception (left) provides a view of a Voyager spacecraft passing behind the rings of Saturn.

ed Saturn in August
d on to an encounter
uary 1986. *Voyager 2*
August 1989, complet-
the gas giants. Passing
n pole to bring it closer
Triton, the spacecraft
liptic plane and out of
he two amazingly suc-
bes provided close-up

views and thousands of images of the oute
planets and their ring systems and moons.

Both of the Voyager spacecraft continue
to operate beyond our solar system. The
are measuring the concentration o
charged particles in an effort to detect the
boundaries between the solar wind and in
terstellar space. The Voyagers are also con
tinuing to study ultraviolet sources among
the stars.

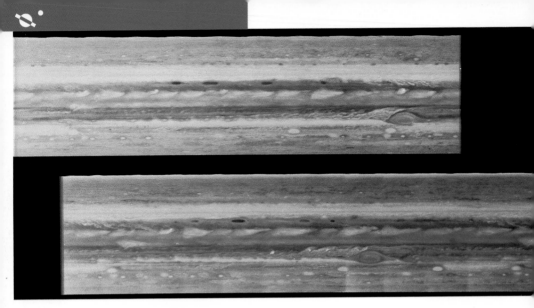

These two images of Jupiter's equatorial region, taken by *Voyager 1* (top) and *Voyager 2* (bottom), reveal the dynamic nature of the Jovian atmosphere. The images are aligned to allow for comparison. In the time between the two Voyager missions, the Great Red Spot moved westward, while the white oval features north of the Spot migrated eastward. The turbulent regions directly to the west and east of the Great Red Spot also show noticeable changes.

Atmosphere

The Jovian atmosphere is thought to consist of hydrogen and about 10 percent helium. It is topped by three layers of clouds. The upper layer, composed of ammonia ice crystals, is the coldest. The temperature near the cloud tops measures about −243°F (−153°C). The middle layer is crystals of ammonium hydrosulfide, a mix of ammonia and hydrogen sulfide. In the lower layer of the atmosphere the temperature and pressure are high enough that clouds composed of water ice or water may exist.

Clouds in Jupiter's atmosphere move in east-west belts and zones that are parallel to the equator. The wind flow within these belts and zones alternates from eastward to westward, and wind speeds vary greatly. Winds are driven by heat from the interior and to a lesser degree by heat from the Sun.

The circulation in the atmosphere is greatly influenced by the planet's rapid rotation. It is thought that the hot gases are deflected into a series of cylinder-shaped columns by the planet's rapid rotation. Along the outer side of the columns, winds flow eastward. On the inner side, they flow westward. These motions are thought to generate the alternating zonal winds carrying clouds with subtle tones of red, blue, brown, yellow, and orange. During the nearly 100 years they have been observed with modern telescopes, these winds have been surprisingly consistent.

Lightning 10,000 times more powerful than any seen on Earth has been detected in Jupiter's clouds. In addition to the extraordinary belts and zones in the Jovian atmosphere, there are a remarkable number of oval features that are thought to be stable eddies, or circular winds. These eddies drift in the zones in which they are trapped by opposing zonal winds. The largest eddy is Jupiter's Great Red Spot.

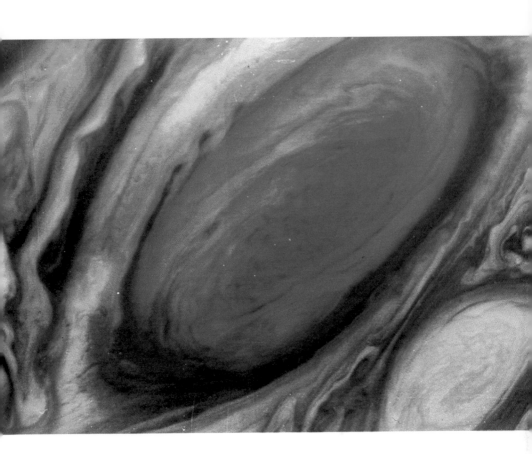

The Great Red Spot

A huge hurricane-like storm system almost as wide as Earth has been observed in Jupiter's southern hemisphere for more than 300 years. Rotating counterclockwise about every six days, the Great Red Spot creates turbulent cloud patterns in its wake. Why the Great Red Spot and other eddies are so long-lived is not completely understood. They must continually absorb energy in order to survive. The energy may come from heat sources below the storms or from the zonal currents in which they are trapped. The energy may also be gained by engulfing and absorbing the energy of smaller eddies.

Perhaps Jupiter's most distinctive atmospheric feature, the Great Red Spot measures about 25,000 miles by 20,000 miles (40,000 km by 32,000 km). This close-up view is from the imaging system of *Voyager 2*.

If the surface of Earth could be peeled off like the skin of an orange, it would fall just short of covering Jupiter's Great Red Spot.

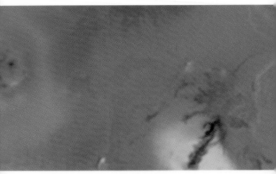

Unlike most moons, Io does not have a surface scarred with impact craters. Io's hundreds of volcanoes have blanketed the surface with sulfurous lava and other ejecta. The prominent volcano (left center) called Pele was one of eight volcanoes erupting plumes of material when *Voyager 1* flew by in 1979.

The Galilean Satellites

Jupiter is orbited by 16 satellites, or moons, four of which are planet-size bodies. All 16 satellites orbit in Jupiter's equatorial plane, creating the appearance of a miniature solar system.

The four largest moons, discovered by Galileo in January 1610, are called the *Galilean satellites* in his honor. In order of distance from Jupiter they are Io, Europa, Ganymede, and Callisto. The average density of these moons decreases with distance, much as the average density decreases from the inner to the outer planets. It was not until the *Voyager* spacecraft visited the Jovian system that the diverse and complex nature of these worlds was discovered.

Io Io is truly one of the most unusual bodies in the solar system. It is about the same size as Earth's Moon. Its average density of 3.5 suggests that, like the terrestrial planets, Io is a rock or silicate-rich body. Yet it is clearly the most volcanically active body in the solar system. Evidence of nine active volcanoes sprouting plumes of material high above Io's surface was discovered in the *Voyager* images. In addition, Io has as many as 200 volcanic craters larger than 12 miles (20 km) across—over 10 times as many volcanoes as Earth has in that size range. About 11 billion tons (10 trillion kg) of volcanic material is erupted yearly—enough to erase all evidence of impact cratering.

The color of Io's surface may come from sulfur or sulfur-bearing compounds. Some speculate that lava flows on Io are made of sulfur and that the shades of yellow and orange and black are various compounds of sulfur that form at different temperatures. Others believe that the flows are similar to those on the terrestrial planets—basaltlike in composition and colored by sulfur contained in and coating the rocks. Some of the lava flows from Io's calderas, or volcanic craters, extend over 100 miles.

Io's volcanoes have low elevations, unlike the large shield volcanoes on Venus, Earth, and Mars. However, the largest volcano on Io, called Pele, is around 870 miles (1,400 km) across—about

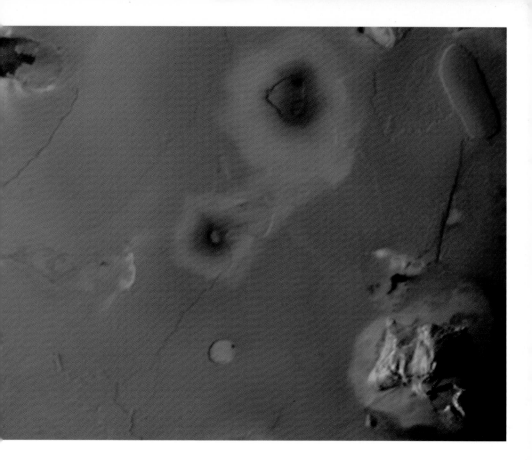

the size of Alaska. Io has mountains that are up to 6 miles (10 km) high, but these are not volcanic. The origin of these mountains is still a mystery.

Volcanic material is ejected from vents on Io at velocities up to 0.6 miles (1 km) per second. The umbrella-like shape of these plumes suggest that Io's volcanoes function much like geysers. High ejection velocity coupled with Io's weak gravitational field allows plumes to soar 180 miles (300 km).

Why is Io so volcanically active? The answer is found in Jupiter. Io orbits at a distance of 262,200 miles (421,600 km)— only about 23,000 miles (37,000 km) farther than the Moon is from Earth. Because Jupiter is over 300 times more massive than Earth, its immense gravity subjects Io to huge tidal forces. These forces create a tidal bulge that moves across Io's surface, causing its crust to flex, or bend, back and forth. This movement generates enough heat to melt the interior and produce Io's widespread hot spot volcanism.

Io's thin atmosphere is mostly sulfur dioxide, which volcanic vents continue to supply. Much of the sulfur dioxide freezes out at night.

Only in the polar regions do mountains, rather than volcanic domes, appear in Io's landscape. Haemus Mons is seen in the lower right of this image of Io's south pole. Also visible are patches of sulfur dioxide frost (bright white) as well as volcanic features such as lava flows.

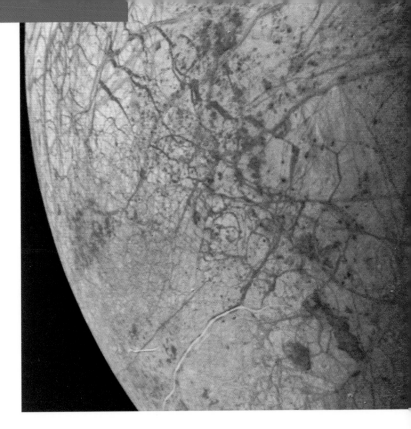

This color image of the satellite Europa, taken by *Voyager 2* in July 1979, clearly shows its smooth but fractured surface. The very small number of impact craters on its surface indicates that Europa has undergone resurfacing processes in the recent past. In February 1995, the Hubble Space Telescope found a thin atmosphere of oxygen on Europa.

Europa At roughly the same size and with nearly the same average density (3.04) as Io, Europa is nonetheless very different. Europa's surface of water-ice is remarkably flat and smooth. The most prominent feature is a network of dark grooves and ridges. The grooves appear to be fractures or cracks in Europa's ice crust that were filled in with water. The ridges are thin and have a maximum elevation of only about 500 to 600 feet.

The relatively high average density of Europa suggests that its ice crust cannot be thicker than 47 to 62 miles (75 to 100 km), suggesting that all of Europa's water now lies frozen on its surface. The absence of impact craters indicates that Europa must have been flooded by water, removing evidence of impact events.

Europa orbits Jupiter at a distance of 417,000 miles (670,900 km). For this reason it is not subjected to tidal forces as powerful as those that affect Io. Still, the tidal forces are strong enough to flex Europa's crust and heat its interior. The heat is released through tectonic activity that cracks and shifts portions of the crust and melts or releases water that floods the surface. Thus, tidal forces may be the source of Europa's grooves and ridges. It has been suggested that tidal heating might be great enough to prevent Europa's crust from completely freezing. If this is true, a layer of water may exist under the ice crust.

Ganymede's icy surface has fewer craters than Callisto's, suggesting that it remained geologically active longer. The brightest areas visible in this *Voyager 2* image of Ganymede are relatively recent craters. Light brown regions within Galileo Regio are probably older impact sites. Galileo Regio also has a large bright area that may be covered with frost.

Ganymede With a diameter of 3,269 miles (5,260 km) and an average density of only 1.93, Ganymede is an ice giant. Roughly half ice and half rock, it is the largest satellite in the solar system, larger even than Mercury. Ganymede orbits Jupiter at a distance of 664,900 miles (1,070,000 km).

Two types of surface on Ganymede are clearly distinguishable by their degree of brightness. Heavily cratered dark regions are the oldest parts of the surface. A large, roughly circular area of dark, heavily cratered terrain about 2,000 miles (3,200 km) across is called Galileo Regio. The only feature in this dark terrain, aside from craters, is a series of parallel furrows. The origins of Galileo Regio and of Marius Regio, another large, circular area of dark terrain, are not clearly understood. They may be the remains of ancient impact basins or evidence of satellite-wide expansion of the crust.

Lighter regions have fewer impact craters, which indicates that they are younger than the dark terrain. They contain tectonic features consisting of parallel curving grooves. These features have no counterparts on the terrestrial planets.

The crust of Ganymede is thought to be about 47 miles (75 km) thick. Beneath lies a mantle of either water or ice, and a rocky or silicate-rich core. There is no question that tectonic forces have reshaped the surface of Ganymede. The system of grooves appears to have formed when the crust stretched or extended, creating fractures and graben-like features. Tectonic forces may have also fractured and rotated large blocks of the dark, heavily cratered terrain and may have faulted and offset parts of the groove systems. Most of the tectonic activity probably occurred early in Ganymede's history, when the interior was still warm. In time, Ganymede's interior cooled, the crust became thicker and stronger, and tectonic activity ceased.

Callisto The outermost of the Galilean satellites, Callisto is nearly Ganymede's twin in size and density. Like Ganymede's, its average density suggests that it is water-rich, with a thick ice crust overlying a mantle of water or ice and a rocky core. It orbits Jupiter at a distance of 1,171,300 miles (1,884,600 km).

Callisto is the most heavily cratered body yet observed in the solar system. Its surface is covered with impact craters. It is the only densely cratered body with no plains between the craters such as those found in the highland regions on Mercury, Mars, and the Moon. The nature of Callisto's water-ice crust is revealed by the shapes of its craters and impact basins, which are much flatter than those on the terrestrial planets. The impact basins also lack the concentric, uplifted mountains and central depressions that characterize these features on the terrestrial planets. Valhalla, at about 1,864 miles (3,000 km) in diameter, is Callisto's largest impact basin. A bright, ringed structure, it was probably hit by an asteroid-size object. The rings, although extensive, have little relief and probably reflect the fracturing that occurred when the ice crust expanded in response to the impact.

Callisto's history of heavy cratering is apparent in this *Voyager 1* image. Unlike Europa, Callisto has not experienced resurfacing for billions of years, so its many impact craters have been preserved. Cratering sites appear bright because impacts expose ice in the crust. The bright area on the left is the giant impact basin Valhalla.

The Lesser Satellites

The other twelve moons of Jupiter are tiny in comparison to the Galilean satellites. Amalthea, the fifth largest, is only 106 miles (170 km) across, but it is the giant of the four small inner satellites that orbit between Jupiter and Io. Metis, Adrastea, and Thebe are only about 16 to 62 miles (25 to 100 km) across.

Four other small satellites are located beyond the Galileans. They range in size from about 9 to 115 miles (15 to 185 km) across, followed by four very different outer satellites. These range from about 19 to 31 miles (30 to 50 km) across and have more elliptical orbits that are more inclined to Jupiter's orbit than the other twelve. More surprising, they orbit Jupiter in the opposite direction to that of its other satellites, and opposite to the direction of Jupiter's spin. These characteristics suggest that these outer satellites are captured asteroids, not originally part of the Jupiter system.

Magnetic Field and Radio Emissions

A powerful magnetic field is generated by currents in Jupiter's metallic liquid hydrogen interior. This magnetic field dominates the behavior of electrically charged particles in a region surrounding the planet called the *magnetosphere*. Jupiter's magnetosphere is enormous. It engulfs seven of its satellites and covers an area much larger than the Sun. It contains radiation belts where charged particles of the solar wind interact with Jupiter's magnetic field.

Periodic bursts of radio energy that come from Jupiter seem to coincide with the movement of Io. They may be generated when the satellites penetrate the radiation belts. In addition, a constant flow of electrons from Io is trapped between the magnetic lines of force that strike Jupiter at about 8° from the poles. This flow is estimated at 3 million amps. This may be the source of the constant background radio emission from Jupiter. In infrared photographs, the areas around both poles and the two spots where Io's electrons are focused glow brightly.

Jupiter's magnetosphere is so huge that, if it were visible, it would be as large as the Moon in the Earth's sky.

If you were there

If you weighed 100 pounds (45 kg) on Earth, you would weigh about 254 pounds (113 kg) on Jupiter.

If you investigated the cloud tops you would find the temperature to be a chilly −243°F (−153°C).

You could be caught in a storm with winds up to 200 miles per hour (320 km/hr)— 3 times as strong as a hurricane on Earth.

The Rings

Pioneer 11 surprised space scientists when its images indicated there might be a ring around Jupiter. *Voyager 2* confirmed that there is actually a system of extremely thin rings, invisible from Earth. The main ring is about 4,350 miles (7,000 km) wide and only about 19 miles (30 km) thick. Its inner edge is about 76,500 miles (123,000 km) above Jupiter's cloud tops. It contains larger objects that may be fragments of small satellites that were broken up by impacts, or the remains of material that never accreted, or-combined, to form a satellite.

A more tenuous inner ring extends 12,430 miles (20,000 km) below the main ring. And an even more tenuous outer ring extends some 52,800 miles (85,000 km) above the main ring.

Very small particles that are only a thousandth of an inch (roughly one micron) in diameter make up the bulk of these rings. Because of their small size, the particles escape from the rings in a relatively short time. Yet the rings remain a permanent feature, meaning that the lost particles are being continually replaced. They are probably created when micrometeorites knock particles off the larger chunks in the main ring. Particles shed by Jupiter's satellites may also become part of the rings.

Voyager 2 captured this image of Jupiter's faint rings from a distance of 900,000 miles (1,450,000 km). Images were collected while the spacecraft was in Jupiter's shadow.

The Shoemaker-Levy Impacts

These comparative images taken by the Hubble Space Telescope show how Shoemaker-Levy impacts changed the surface of Jupiter. The image at left was taken in May 1994, two months before impact. The image above, taken in July 1994, shows several impact scars.

In March 1993, an elongated object was discovered in the vicinity of Jupiter. It was determined that this "object" was a string of 21 fragments of a comet which was broken apart by tidal forces as it passed by Jupiter. The fragments of the comet, named Shoemaker-Levy after its discoverers, were on a collision course with Jupiter. If the largest of these fragments had struck Earth, it is estimated that it would have made a crater the size of Rhode Island.

In July 1994, Shoemaker-Levy slammed into Jupiter's atmosphere at an estimated speed of 130,000 miles per hour (210,000 km/hr). The impacts lasted almost a week. Some of the impacts created plumes or fireballs that were up to 2,500 miles (4,000 km) wide and reached heights up to 1,300 miles (2,100 km). These plumes flattened out and fell back on the cloud tops of Jupiter's atmosphere.

Clearly visible markings were left in Jupiter's atmosphere after each collision. The largest of these is almost as large as Earth. One theory is that these dark patches are the result of carbon-bearing materials deposited in Jupiter's atmosphere by the comet fragments. It was hoped that the fragments might penetrate deep enough to expose the chemical components in cloud layers in Jupiter's upper atmosphere, but this does not appear to have occurred. Ammonia and sulfur were detected, but the sulfur may have come from the comet fragments.

Saturn: The Ringed Giant

The second largest of the gas giants, Saturn possesses a feature unique in the solar system—its spectacular rings. Saturn is about 85 percent the size of Jupiter but twice as far from Earth, which makes Earth-based observation difficult. As with Jupiter, highly detailed images and data had to await exploration of Saturn by robotic probes. These have revealed the complex nature of its rings, high-velocity winds in its atmosphere, and new information about its 18 moons, the largest and most mysterious of which is Titan.

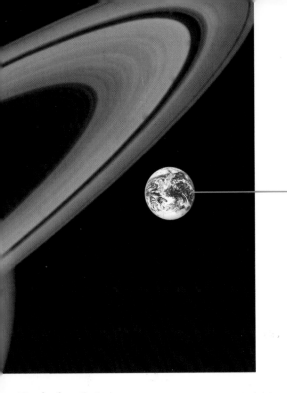

Earth

Exploring Saturn

On September 1, 1979, the *Pioneer 11* probe passed within 13,000 miles (21,000 km) of Saturn and sent back the first close-up images of the planet and its rings. However, the images revealed little new information about Saturn's cloud patterns and swirling atmosphere.

Pioneer 11 was followed by the more sophisticated *Voyager* probes, both launched in 1977. In November 1980, about a year and a half after the *Voyagers'* encounters with Jupiter, *Voyager 1* passed within 77,143 miles (124,123 km) of Saturn and then moved out of the solar system. *Voyager 2* passed within 62,980 miles (101,335 km) of the planet in August, 1981 and then proceeded on to Uranus and Neptune. The *Voyagers* provided an incredible wealth of images and data, which have greatly increased our understanding of the Saturnian system.

NASA and the European Space Agency (ESA) are cooperating on a mission to the Saturnian system called *Cassini*. Like the *Magellan* spacecraft which imaged Venus, *Cassini* will use radar to image the surface of Saturn's moon Titan through its cloud cover. An entry probe will analyze the chemistry of Titan's atmosphere during a nearly three-hour descent and will continue its analysis from the surface of Titan if the probe survives. The mission is planned for 1997 and the data it sends back should be received early in the next century.

Voyager 2 created this color composite image as it approached Saturn in 1981. At a distance of 13 million miles (21 million km), the planet's light and dark atmospheric bands are clearly visible.

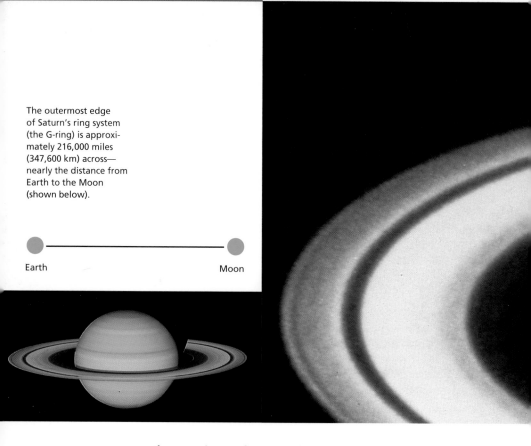

The outermost edge of Saturn's ring system (the G-ring) is approximately 216,000 miles (347,600 km) across—nearly the distance from Earth to the Moon (shown below).

Earth

Moon

The Interior and Atmosphere

Like Jupiter, Saturn is composed mostly of hydrogen and helium, although it has only about 3 percent helium in the atmosphere, less than half of Jupiter's 10 percent. Saturn's mass is about 95 times greater than Earth's but it has 800 times the volume. Its average density is 0.7, the lowest of any planet in the solar system. In fact, its density is so low that, unlike the other planets, it would float if it were placed in a pool of water.

The interior structure of Saturn, like Jupiter's, can be deduced from its shape and gravitational field. Saturn's rotation of 10 hours, 40 minutes makes it flatten at the poles and bulge at the equator more than any other gas giant. Its equatorial radius is 3,700 miles (5,900 km) greater than its polar radius. This suggests that Saturn, like Jupiter, has a dense core of melted rock or rock-ice mixtures about the size of 10 to 15 Earth masses. The core is surrounded by a mantle of metallic liquid hydrogen. Saturn's mantle is surrounded by ordinary liquid hydrogen.

With decreasing temperature and pressure, the liquid hydrogen becomes gaseous. At the cloud tops, the temperature is about

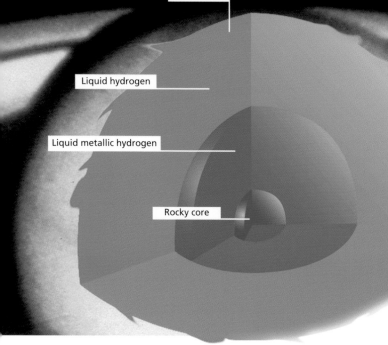

Atmosphere

Liquid hydrogen

Liquid metallic hydrogen

Rocky core

−301°F (−185°C). In the outer core, in contrast, the temperature reaches 21,700°F (12,000°C) and the pressure reaches 12 million times the pressure on Earth's surface.

Saturn's atmosphere is thought to consist of three cloud layers, much like Jupiter's, but it is colder and less complex. The bottom cloud layer of water ice crystals or droplets is three times deeper in the atmosphere than Jupiter's comparable layer of clouds. Clouds of ammonium hydrosulfide form the middle layer, and the topmost layer consists of ammonia ice crystals.

Saturn's yellow clouds move in zones parallel to the equator, with winds that may alternate from eastward to westward between zones. Jupiter has more of these zones than Saturn, but the wind speeds in Saturn's zones are much higher. Speeds as high as 1,100 miles per hour (1,800 km per hour) have been measured in the broad equatorial zone, which is more than 50,000 miles (80,000 km) wide and extends from 40° north of the equator to 40° south of it. This is more than ten times hurricane force on Earth and four times the wind speed on Jupiter.

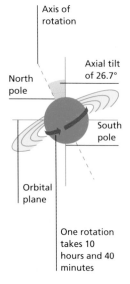

Axis of rotation

Axial tilt of 26.7°

North pole

South pole

Orbital plane

One rotation takes 10 hours and 40 minutes

Because Saturn is so far from the Sun, its atmosphere receives only a little over 1 percent of the solar energy Earth receives. However, like Jupiter, Saturn radiates about twice as much heat as it absorbs from the Sun. Thus the circulation pattern in the atmosphere is generated by gases in convective motion driven by interior heat and influenced by Saturn's rapid rotation.

Saturn has fewer hurricane-like storms than Jupiter. There is a red spot in Saturn's southern hemisphere, but it is only one-third the size of Jupiter's Great Red Spot.

The Rings

When Galileo pointed his newly made telescope at Saturn, he observed strange projections, or bulges, on the sides of the planet that he could not explain. In 1610, when he reported his observations, he thought he had found separate bodies on both sides of Saturn. More puzzling yet, in 1612 the protuberances he had dubbed "ears" seemed to disappear. No one understood the nature of this phenomenon until 1659, when Christiaan Huygens saw that the protuberances were parts of a ring that did not touch the planet. When Earth is directly in the plane of the rings, they are seen edge-on and seem to disappear. Using the long-focus telescope that he had designed, Huygens also spotted Titan. In 1675 Italian astronomer Jean-Dominique Cassini, who was then supervisor of the Paris Observatory, discovered the gap in the rings that bears his name.

In September 1990, about five months after it was launched, the Hubble Space Telescope took this image of a huge white spot that spread into a series of clouds moving in Saturn's equatorial zone.

Composition of the Rings With the telescopes available in Cassini's time, Saturn's rings looked so opaque that they were thought to be solid. We now know that they consist of countless particles with an average density near 1, which are thought to be composed of water ice and dust, similar to dirty snowballs.

The particles appear to vary greatly in size. Most are no larger than an inch across, but the rings are also believed to contain clumps of particles that range from several feet to over a half mile across. For every very large object in the rings there are many more medium-sized objects, and so forth, down to the smallest particles. If all the particles in Saturn's rings were collected to form a single satellite, it would be about 60 miles (100 km) wide.

The thickness of Saturn's main ring system ranges from only about 33 to 330 feet (10 to 100 m), but it has a width of about 28,000 miles (45,000 km). The entire ring system is estimated to be about 250,000 miles (400,000 km) wide.

The system consists of several clearly observable rings of differing brightness, some with gaps between them. *Voyager* images revealed that the major rings consist of many tiny rings—about 1,000 have been identified. Some "ringlets" appear to be braided together. Above the main or B-ring, dark radial "spokes" have been observed appearing and disappearing. It is thought that when Saturn's magnetic field electrostatically charges dust particles, they temporarily rise above the ring, creating the spokes.

This composite of Saturn's northern hemisphere was assembled from ultraviolet, violet and green images obtained in August 1981 by *Voyager 2*, from a range of 4.4 million miles (7.1 million km). The cloud patterns evident include three spots moving westward at about 33 miles per hour (53 km/hr).The ribbonlike feature to the north marks a high-speed jet where wind speeds approach 330 miles per hour (530 km/hr).

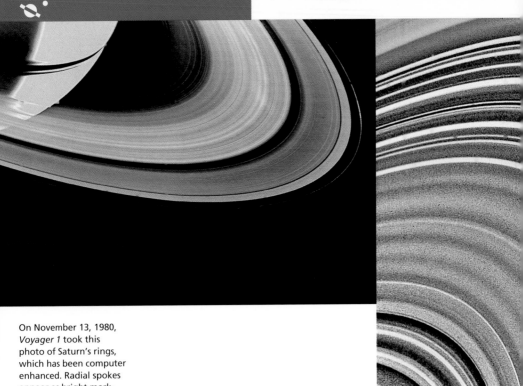

On November 13, 1980, *Voyager 1* took this photo of Saturn's rings, which has been computer enhanced. Radial spokes appear as bright markings in the B-ring.

Origin of the Rings Saturn's rings developed from a cloud of particles that came from the breakup of a satellite or from material that did not accrete, or combine, into one or more moons. For the most part, the rings are located within what is known as the *Roche limits* of the planet. Named after the French mathematician, Edouard Roche, who proposed the theory in 1850, this refers to the region within which tidal forces will tear a satellite apart. In this cloud of debris, particles in different orbitals collided. This reduced their velocities. Over time, the cloud of orbiting, colliding particles tended to flatten into the plane of the equator, forming a ring about the planet.

Particles within the flattened disk that are close to the planet move faster than those farther out. The interaction between faster- and slower-moving ring particles tends to push the faster-moving ones inward and the slower-moving ones outward. This causes the ring to spread out.

The presence of satellites both inside and outside the rings greatly influences their structure. The gravitational pull of the satellites affects the orbits of the particles, causing gaps in the rings. These are called *shepherd satellites* because they confine the particles to narrow rings. Satellites embedded in the rings are also a source of new particles.

This image of Saturn's rings was made by combining three *Voyager* pictures taken with separate ultraviolet, blue, and green filters. Special processing brings out color differences between the rings, which indicates the composition of the particles in each ring. The blue C-ring, which takes up most of the image, has several ringlets that appear yellow, similar to the B-ring at the top and right of the image.

If you were there

If you weighed 100 pounds (45 kg) on Earth, you would weigh 93 pounds (42 kg) on Saturn.

Each season lasts more than 7 Earth years.

If you could place it in a water-filled tub large enough to contain it, Saturn would float.

The Ring System The ring system of Saturn is identified by letter names, generally according to the order in which the rings were discovered. Closest to Saturn is the faint D-ring. Its inner edge is about 4,200 miles (6,700 km) from the cloud tops. It extends about 4,700 miles (7,500 km) out to where the more visible but tenuous C-ring begins. The C-ring has a width of about 11,000 miles (17,500 km), placing its outer edge about 37,500 miles (60,000 km) above the cloud tops.

Next is the bright B-ring. Thought to be the densest of the rings, it is opaque enough to cast a shadow on Saturn. The B-ring is about 16,000 miles (25,500 km) wide, with its outer edge marking one of the boundaries of the Cassini Division. First observed in 1675, the Cassini Division is a gap 2,900 miles (4,700 km) wide that separates the B-ring from the A-ring. The gap probably formed as a result of the gravitational interaction between ring particles and Saturn's moon, Mimas.

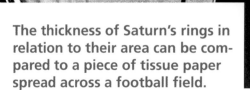

The thickness of Saturn's rings in relation to their area can be compared to a piece of tissue paper spread across a football field.

Voyager missions revealed for the first time the astonishing complexity of Saturn's rings. The *Voyager* image of Saturn's A-ring (left) was taken from a distance of about 1.7 million miles (2.8 million km). Colors have been enhanced to show differences between the rings. The close-up view of the B-ring, (right), detailing areas as small as 6 miles (10 km) across, displays the hundreds of ringlets contained within a segment of the ring.

The A-ring is 9,100 miles (14,600 km) wide, and its outer edge is about 48,000 miles (77,000 km) above the cloud tops. Between 20 and 310 miles (30 and 500 km) wide, the peculiar F-ring is very narrow. It is located about 50,000 miles (80,000 km) above the cloud tops. Particles within the F-ring are kept in place by the small shepherd satellites, Prometheus and Pandora.

Beyond the F-ring is the tenuous G-ring, discovered by the *Voyager* spacecraft, and the even more tenuous E-ring. The indistinguishable outer edge of the E-ring is estimated to lie somewhere around 260,000 miles (420,000 km) above the cloud tops.

Magnetic Field and Magnetosphere

As on Jupiter, the currents in Saturn's mantle of liquid metallic hydrogen generate a strong magnetic field. Saturn's field is 36 times less powerful than Jupiter's but 540 times more powerful than Earth's. The axes of most planetary magnetic fields do not coincide with their rotational axes. The separation between the two on Earth is 11.5°. Saturn is the exception with a separation of less than 1°.

How Did *Voyager* Analyze the Rings?

As *Voyager 1* flew behind Saturn and its rings, its radio signal passed through the planet's rings before it was received on Earth. The varying strength of the radio signal was measured to help determine the size of the particles within the rings. Smaller particles went undetected, but the measurement revealed that some of the particles in the rings were between 6 feet (2 m) and 25 feet (8 m) in diameter.

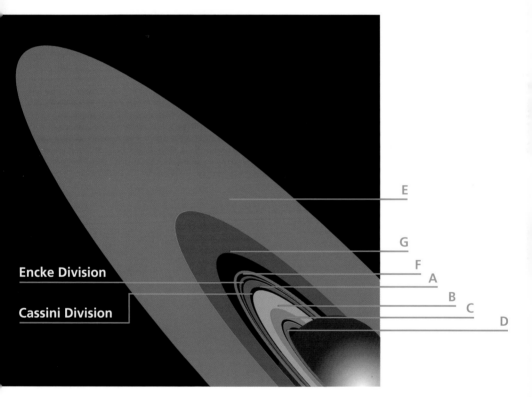

E

G

F

Encke Division

A

B

Cassini Division

C

D

Saturn's magnetosphere, the region where the magnetic field controls the activity of charged particles, is intermediate in size between Earth's and Jupiter's. It extends beyond the orbit of Titan. Fewer particles are trapped in Saturn's magnetosphere than in Jupiter's. Saturn's extensive ring and satellite systems are efficient absorbers of charged particles, creating gaps in the radiation belts. The materials that make up the A-ring are so efficient that the region is nearly devoid of charged particles.

Cassini Mission

The *Cassini* orbiter will be launched in October 1997 and is scheduled to arrive at the Saturnian system in June 2004. *Cassini* will release a probe through the atmosphere of Titan. Since its atmosphere is known to contain complex organic molecules, Titan has been compared to the primordial Earth. The *Huygens* probe will gather data about Titan. Meanwhile, *Cassini* will spend at least four years orbiting Saturn. It will make dozens of flybys of Saturn's many moons, study the polar and equatorial regions of the planet, and explore the mysteries of the rings.

The surface of Titan, Saturn's largest moon, is hidden by its dense clouds. However, this computer-enhanced image of a view from *Voyager 2* captured the beauty of the Titanian upper atmosphere.

Titan: The Giant Moon

A planet-size body larger than Mercury, Titan has a diameter of 3,190 miles (5,150 km). It is the second largest satellite in the solar system, exceeded only by Jupiter's Ganymede. It orbits Saturn at a distance of about 760,000 miles (1,222,000 km).

Titan has an average density of 1.88, almost the same as Ganymede's and Callisto's. Like the large moons of Jupiter, Titan is thought to be about half water ice and half rock or silicates. Its rocky core is surrounded by a mantle of ice, and beneath its ice crust there may be a layer of liquid water.

In spite of their similarities, Titan has an atmosphere, while Ganymede and Callisto do not. The reason may be that the temperature in the Saturnian system was much lower than in the Jovian system. Temperatures in the region around Saturn did not become high enough to drive off the methane, ammonia, and nitrogen trapped in ices, so those gases were not released until Titan formed. This created an early atmosphere that was later added to by outgassing from the interior.

Titan's atmosphere contains orange clouds that completely obscure its surface. The atmosphere is over 80 percent nitrogen. It also contains methane, ethane, argon, and hydrogen. All the oxygen is tied up in water ice. The atmosphere is four times denser than Earth's. However, the gravitational pull is weaker, so the atmospheric pressure is only 1.6 times greater than Earth's.

The temperature on the surface of Titan is far below the freezing point of water but near the freezing point of methane. Below

Titan: Images from Hubble

The first images of large-scale surface features on Titan's surface were taken in October 1994 using the Hubble Space Telescope's Wide Field Planetary Camera 2. These images were taken at near-infrared wavelengths where the atmospheric haze is transparent enough to map the reflectivity of surface features. The four images above are a composite of a series that were taken of the satellite's hemispheres over a perio[d] about 16 days (one rotation of Titan).

Scientists speculating on the natu[re] the bright and dark areas in the im[age] think they confirm the existence of s[olid] solid surface. The most prominent b[right] area, shown in the top right image, is [esti]mated to be a surface feature about 2[,500] miles (4,023 km) across—about the si[ze of] the continent of Australia.

clouds of methane, liquid hydrocarbons in the form of ethane may have accumulated to form lakes or even oceans. Hydrocarbons and nitrogen react to form carbon-nitrogen compounds. This mix of complex organic compounds may coat Titan's surface with a brownish-orange tarlike substance. On Earth, such compounds wash into the ocean, where the first life forms developed. The Hubble Space Telescope recently returned images of bright and dark areas that may be impact craters, oceans, continents, or other features yet to be identified.

Other Satellites of Saturn

Having formed in a volatile-rich and colder environment than the Jovian satellites, Saturn's satellites retained water, methane, ammonia, and nitrogen that condensed from the solar nebula. Not heavily modified, their surfaces preserve their early histories.

About one-quarter of Saturn's moon Mimas (left) is dominated by a massive crater, estimated to be 60 miles (100 km) wide. The high-resolution composite of Saturn's moon Enceladus (right) was obtained by *Voyager 2* in August 1981—from a range of 74,000 miles (119,000 km).

Unlike Jupiter, Saturn has only one planet-size moon. Titan is more than three times larger than the next largest satellite of Saturn. Excluding Titan, the six largest satellites range in diameter from 240 to 950 miles (390 to 1,530 km). In order of increasing distance from Saturn, these are Mimas, Enceladus, Tethys, Dione, Rhea, and Iapetus. These bodies range in average density from 1.1 to 1.4, indicating that their interiors are at least 50 percent water ice.

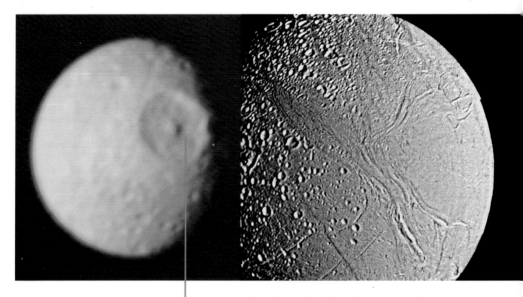

Herschel, the giant crater on Mimas

Viewing the first images of Mimas transmitted by *Voyager 1*, planetary scientists were struck by its resemblance to the fictitious Death Star in the film *Star Wars*.

Mimas The nearest of the Saturnian satellites, Mimas has a diameter of 240 miles (390 km). Its heavily cratered surface could be mistaken for the highlands of the Moon. It has a large impact crater named Herschel, which is nearly one-third the diameter of the entire satellite. The impact that formed this crater is about the largest Mimas could have withstood without shattering.

Enceladus Next is Enceladus, an icy white satellite, the most reflective of any known planetary body. It is 311 miles (500 km) wide. Its surface is heavily cratered in one hemisphere and relatively smooth and craterless in the other. Narrow linear grooves suggest that the ice crust has been fractured, which may have allowed water from the interior in the form of icy flows to modify a heavily cratered surface.

Tethys With a diameter of 659 miles (1,060 km), Tethys has a heavily cratered surface, like Mimas. Tethys also has a huge impact crater called Odysseus that is proportional in size to Herschel on Mimas. Like Enceladus, Tethys has some smooth regions with fewer impact craters. This indicates that water may have poured out onto its surface and covered some craters. The presence of a valley system called Ithaca Chasma suggests tectonic activity. This system, extending three-quarters of the way around the satellite, is more than ten miles wide and may have formed by fracturing and rifting of the ice crust.

On Saturn's ice-covered moon, Tethys, the Ithaca Chasma valley stretches 60 miles (97 km) wide and 2 miles (3 km) deep, about three-quarters of the way around the satellite.

Odysseus, an impact crater on Tethys

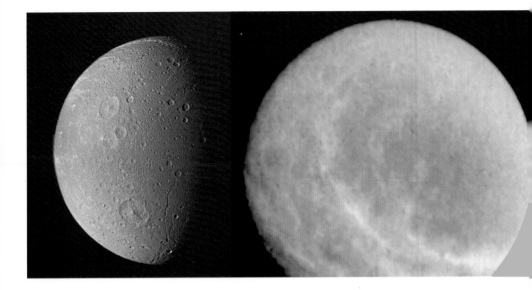

Many impact craters, the results of the collision of cosmic debris, are shown in this *Voyager 1* color mosaic of Saturn's moon Dione (left). The images were taken from a range of 100,600 miles (162,000 km) in late 1980. Because the surface of Saturn's moon, Rhea (right), is mostly ice, it is highly reflective, and almost uniformly white in appearance.

Dione Dione is similar to Tethys in size and surface features. It has a diameter of 700 miles (1,120 km) and its surface has regions dominated by craters and other nearly craterless regions. Like Tethys, Dione has branching valleys that are more than ten miles wide and hundreds of miles long. As with Tethys and Enceladus, fracturing of the ice crust may have allowed ice flows to cover or resurface some regions. Wispy streaks seen on Dione may be ice deposits from water that flowed into surface cracks.

Rhea Saturn's second largest satellite, Rhea is 950 miles (1,530 km) wide. Parts of its surface are heavily cratered and others are not, again suggesting possible resurfacing by icy flows. Like Dione, Rhea has wispy white streaks that crisscross the surface in one hemisphere and that may have formed ice-filled cracks in the crust.

Iapetus At 908 miles (1,460 km) wide, Iapetus is nearly the same size as Rhea but is much more distant from Saturn. One hemisphere is dominated by craters, but the other hemisphere is unique. It is 10 times darker than the brighter, cratered hemisphere, apparently because its surface is coated with a dark material whose origin is still not fully understood. It may be a thin layer of organic material that came from an external source, such as a comet. However, the presence of some dark materials in the floors of craters in the opposite hemisphere suggests it may come from the interior of Iapetus.

This mosaic of Saturn's moon, Iapetus, was composed from images taken by *Voyager 2* in August 1981 at a range of 680,000 miles (1.1 million km). The northern hemisphere is shown and craters may be seen in both the bright and dark regions.

Minor Satellites of Saturn

Saturn has at least eleven smaller satellites, many of which are irregularly shaped. They range in diameter from 12 to 159 miles (20 to 255 km). These are Pan, Atlas, Prometheus, Pandora, Epimetheus, Janus, Telesto, Calypso, Helene, Hyperion, and Phoebe. Their shapes, and the unusual nature of some of their orbits, suggest that many of these satellites are fragments of larger bodies. A few of these unusual icy satellites are reviewed here.

Hyperion Rough and irregularly shaped, Hyperion is the largest of Saturn's minor satellites. The remnants of a crater 62 miles (100 km) wide dominate its surface. Hyperion is probably the remains of a large satellite that was shattered by the impact that caused the crater.

Phoebe The satellite farthest from Saturn, Phoebe, orbits in the direction opposite to the orbits of the other satellites and opposite to the direction of Saturn's rotation. Its orbit is highly inclined. Phoebe is about 137 miles (220 km) wide. Its surface, like that of Iapetus, is coated with a dark, possibly organic material similar to the material that covers the nuclei of comets. It is thought that Phoebe may be a captured nucleus of a large comet.

Janus and Epimetheus The jagged and irregular Janus and Epimetheus are nearly the same size, at 118 miles (190 km) and 75 miles (120 km), respectively. They are called *co-orbitals* because they share very nearly the same orbit. Their orbits are separated by only 30 miles (50 km). One makes the trip in 16.66 hours while the other takes 16.67 hours. This slight difference creates an interesting relationship. The faster satellite catches up with the slower one every four years. Then, instead of colliding, they exchange orbits. The outer, slower one slips into the inner orbit and speeds up. The inner, faster one moves into the outer orbit and slows down. Four years later they meet and switch orbits again. Their similar sizes and co-orbital nature suggest that Janus and Epimetheus are fragments, possibly halves of a destroyed satellite.

About four days after *Voyager 2's* closest approach to Saturn its cameras looked back on the ringed planet to acquire this computer-enhanced view.

Lagrangian Satellites In the eighteenth century, mathematicians calculated that a small object could share an orbit with a large one if it either led or trailed the large object by about 60°. These two points in the larger object's orbit were called *Lagrangian points*. Asteroids were found at these points in Jupiter's orbit around the Sun, and early in 1980 a small satellite of Saturn—Helene—was found at a Lagrangian point in Dione's orbit. Telesto and Calypso were later found at the two Lagrangian points in the orbit of Tethy's.

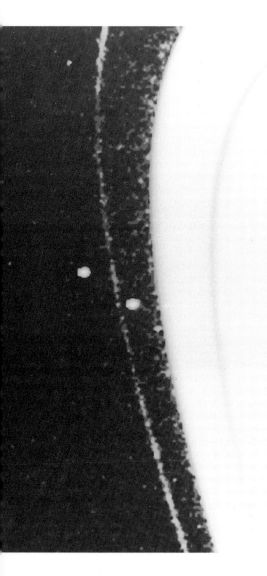

Shepherd Satellites

Shepherd satellites perform their function through gravitational attraction. A satellite orbiting on the outer edge of a ring—and moving more slowly than particles in the ring because of its greater distance from the planet—attracts particles as they go by. Slowed by the pull of the satellite, the particles lose energy and fall into an orbit that is closer to the planet. Thus an outer limit for the ring is set by the position of the outside shepherd satellite. Conversely, a satellite orbiting just inside a ring moves faster than the ring and gives the particles it attracts a boost in energy, causing them to move into a slightly higher orbit. This interaction sets the inner boundary of the ring. A *Voyager 2* image (left) shows Saturn's thin F-ring bracketed by two shepherd satellites.

This true color image of Uranus was taken by *Voyager 2's* camera in January 1986 when the spacecraft was seven days from its closest approach and 5.7 million miles (9.1 million km) from the planet.

Uranus

The gas giants fall into two groups based on their size. Uranus and Neptune are the smaller of the giants. The diameter of Uranus is only about one-third the size of Jupiter's, but four times greater than Earth's. In fact, Uranus is large enough to hold about 64 Earths. With an average density of only 1.3, however, Uranus has only 14.5 times the mass of Earth. Orbiting twice as far from the Sun as Saturn, the dimly lit Uranian system consists of a series of thin rings and 15 moons.

Inspired by the discovery of Uranus, Martin Klaproth, a German chemist, changed the name of his newly discovered element from *klaprothium* to uranium.

Naming Uranus

When William Herschel announced his discovery of a new planet, he named it *Georgium Sidus* (Georgian Star) in honor of George III, the King of England. Others suggested naming it Herschel, but German astronomer Johann Bode proposed that, like the other planets, it be named after a classical god. Starting with Mars, each planet was named after the father of the one before: Jupiter was the father of Mars, and Saturn was the father of Jupiter. Bode proposed the name Uranus for the new planet, after the father of Saturn, who was both the most ancient of the Greek gods and the god of the starry sky.

Earth

Discovering Uranus

In March 1781, the British astronomer William Herschel made a discovery while viewing the skies through one of his telescopes. The bluish-green object he found did not appear as a point, as a star would, but as a small disk. First thinking he had found a comet, he determined that its orbit was nearly circular and beyond Saturn. Herschel had discovered Uranus. With an orbit extending 19 times farther from the Sun than Earth's orbit, Uranus is so distant that its discovery doubled the size of the known solar system. Herschel also discovered the two largest satellites of Uranus, named Oberon and Titania.

Voyager 2 at Uranus Because it is so far from the Sun and because it receives only three-tenths of a percent of the sunlight that falls on Earth, little was known about Uranus until it was visited by *Voyager 2*. In 1986, five years after having flown by Saturn, *Voyager 2* encountered Uranus, passing within about 51,000 miles (82,000 km) of the planet.

Voyager 1 was about 600,000 miles (1 million km) from Uranus, on its way to Neptune, when it recorded this wide-angle view of the planet's crescent in January, 1986.

Unusual Inclinations

Except for Uranus, each planet's axis of rotation is nearly perpendicular to the plane of the solar system (even Venus, which is nearly upside down). Earth's axis, for example, is tilted or inclined 23.5° to its orbital plane. Uranus's axis of rotation is inclined by almost 98°. This means that the rotation axis is 8° below the orbital plane, so the whole Uranian system appears to be tipped over. Consequently, at different times during its 84-year orbital period, we view one of the poles of the planet and then its equator head-on.

Interior and Atmosphere

In contrast to the larger gas giants, the composition of Uranus is not dominated by hydrogen and helium. Hydrogen accounts for only about 15 percent of the planet's mass. Methane, ammonia, and water contribute a greater percentage of the planet's mass, and the interior structure must reflect this difference in composition. Like Jupiter and Saturn, Uranus is thought to have a dense core of melted rock and ice.

The mantle of Uranus is probably hot liquid hydrogen with only a small amount of liquid metallic hydrogen deep in the interior. An ocean composed mostly of water with some methane and ammonia may lie at the base of the atmosphere.

The atmosphere is composed mainly of hydrogen and helium with some methane. As with Jupiter and Saturn, three distinct cloud layers are thought to exist. Ammonia clouds form the top layer, then a layer of ammonium hydrosulfide clouds, and then clouds of water ice. These cloud layers, particularly the water ice clouds, would be found deep in Uranus's atmosphere where temperatures and pressures are higher.

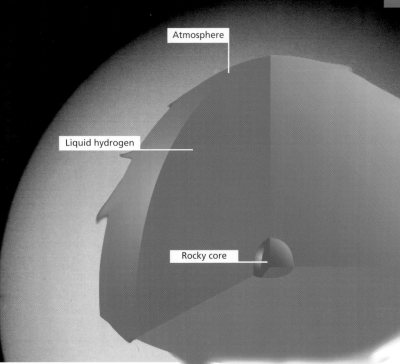

Atmosphere

Liquid hydrogen

Rocky core

The temperature in the upper atmosphere is so cold that methane condenses and forms the thin layer of clouds that lies above the other layers. These methane clouds give the planet its blue-green color. Although no distinct zones and belts or hurricane-like storm systems have been seen, there are strong winds. The equatorial winds move more slowly on Uranus than on the other gas giants, but mid-latitude winds have been measured as high as 375 miles per hour (600 km per hour). The winds are greatly influenced by the rapid rotation of Uranus (17 hours 14 minutes), and they blow mainly eastward parallel to the equator.

Magnetic Field

Uranus generates a strong magnetic field—about 48 times greater than Earth's. Like its rotational axis, the magnetic axis of Uranus has an unusual orientation. It is tilted almost 59° away from its rotational axis. Just as unusual, the center of the magnetic field does not coincide with the center of the planet. It is offset by almost one-third of Uranus's radius—nearly 4,800 miles (7,700 km).

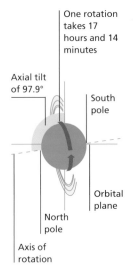

One rotation takes 17 hours and 14 minutes

Axial tilt of 97.9°

South pole

Orbital plane

North pole

Axis of rotation

Rings of Uranus

In 1977 astronomers were observing an occultation (the movement of one celestial body in front of another, blocking the other from view) of a star by Uranus. Before the star disappeared behind Uranus, its brightness dropped in distinct pulses. This was repeated when the star reappeared on the other side of the planet. These pulses were caused by a ring system around Uranus. In 1986, *Voyager 2* confirmed the existence of a ring system when it observed eleven narrow rings. The innermost ring is about 7,700 miles (12,400 km) above Uranus, and the ring system extends out to about 16,200 miles (26,000 km). Most of the rings are no more than 6 miles (10 km) wide. The outermost ring, however, is about 62 miles (100 km) wide. The dark ring particles range from 4 inches (10 cm) to 33 feet (10 m), with meter-size objects the most numerous. The thickness of the rings is comparable to Saturn's rings, ranging from 33 to 330 feet (10 to 100 m).

Taken from the far side of the planet, with back-lighting from the Sun, *Voyager 2* images revealed that the entire ring area contains finer particles organized in tiny ringlets. Particles are continually being replaced, partly by larger objects within the rings or from the surrounding satellites. The outer ring is the most massive, and its particles are kept in place by the only shepherding satellites found—Cordelia and Ophelia.

Pictures shot as *Voyager 2* cameras looked back toward the Sun clearly indicated the presence of dust particles in the rings of Uranus.

A 12-mile (19-km) cliff on Earth similar to the one on Miranda would reach the orbit of the Space Shuttle.

Satellites

Before *Voyager 2*'s encounter, five moons of Uranus had been identified. They range in size from 301 to 963 miles (485 to 1,550 km) across, very similar to the size range of Saturn's largest moons, excluding Titan. *Voyager 2* discovered ten more inside the orbits of the larger moons, bringing the total to 15. These range in diameter from 16 to 96 miles (25 to 155 km).

The largest moons, in order of increasing distance from Uranus, are Miranda, Ariel, Umbriel, Titania, and Oberon. Their average densities are higher than expected, ranging from 1.4 to 1.7. This suggests that they may be more than 50 percent rock or silicates, with smaller percentages of water ice than Saturn's similar-sized moons contain.

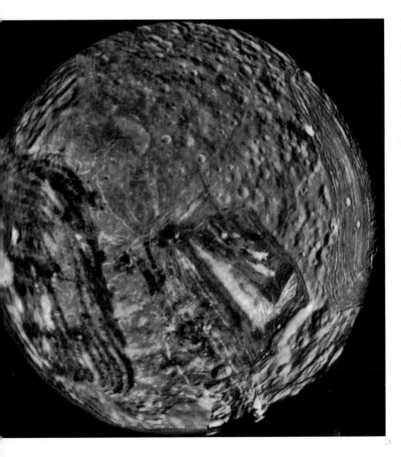

Voyager 2 passed within 18,600 miles (30,000 km) of Miranda, closer than any other object in the Uranian system. In addition to old cratered terrain, Miranda's surface has prominent ovoidal features made up of concentric sets of ridges and grooves.

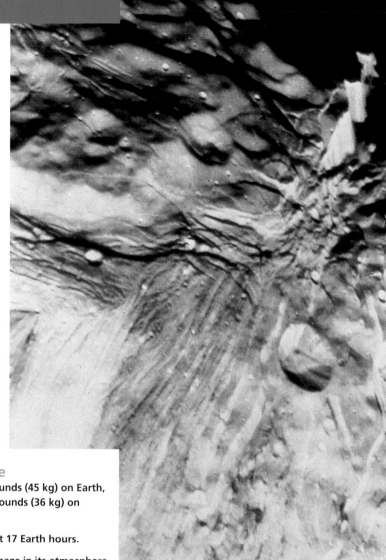

Uranus's moon, Miranda, is seen at close range—from *Voyager 2* images taken on January 24, 1986, at a distance of 22,000 miles (36,000 km). Two distinct terrain types are visible—a rugged, higher-elevation terrain, right, and a striated, lower terrain. The highest point of the cliff is about 9 miles (15 km) above the valley floor.

Miranda One of the most unusual satellites of Uranus, Miranda has a surface like no other in the solar system. Three large features ranging from 125 to 185 miles (200 to 300 km) across dominate the otherwise heavily cratered surface. These features consist of concentric sets of circles and grooves. The reflective properties of the material in them alternate between bright and dark. Another striking landform is a vertical cliff 9 miles (15 km) high. Called Verona Rupes, it appears to be a fault scarp that is without equal on the terrestrial planets. These unusual features may have formed as Miranda reassembled after impacts broke it up, or they may reflect tectonic activity.

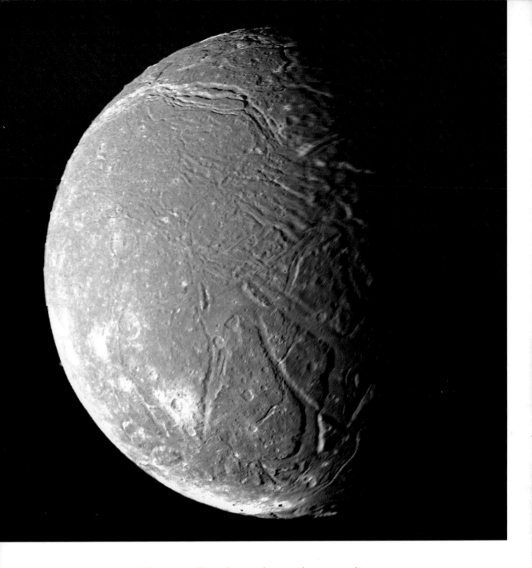

Ariel and Umbriel These satellites have almost the same diameter—721 miles (1,160 km) and 740 miles (1,190 km), respectively. Their surfaces, however, are markedly different. Umbriel's surface is heavily cratered and uniformly dark except for one bright-rimmed crater about 70 miles (110 km) wide. Ariel is far less cratered, and its surface is cut by numerous valleys resembling the valleys on some of Saturn's satellites.

Titania and Oberon These satellites are also similar in size. Titania, the largest in the Uranian system, has a diameter of 1,001 miles (1,610 km), and Oberon has a diameter of 964 miles (1,550 km). Titania, like Ariel, has a less cratered surface cut by fault valleys, which extend over 620 miles (1,000 km). The valleys indicate that the satellite's crust has expanded, causing fracturing that allowed resurfacing by icy flows from the interior.

This mosaic of Uranus's moon, Ariel, is made up of images taken at a distance of 80,000 miles (130,000 km) on January 24, 1986. Much of Ariel's surface is densely pitted with craters 3 to 6 miles (5–10 km) across.

This contrast-enhanced image of Neptune was captured by *Voyager 2* at a range of 9.2 million miles (14.8 million km) in August 1989. The Great Dark Spot, prominent on the right in this picture, was missing from Hubble's 1994 images.

Neptune

Neptune is the fourth largest planet in the solar system and the smallest gas giant. Its diameter is only 990 miles (1,590 km) less than Uranus's and about 3.9 times greater than Earth's. With a density of 1.6, its mass exceeds Uranus's and is 17 times greater than Earth's. Neptune is about 2.8 billion miles (4.5 billion km) from the Sun, about 1.5 times more distant than Uranus. It receives only one-tenth of one percent of the sunlight that reaches Earth. As with all outer planets, most facts known about Neptune were revealed by *Voyager 2*. One of the imaging team members working on that mission compared Neptune's light to the inside of an unlit cathedral on a cloudy day.

Discovery and Exploration

A few years after the discovery of Uranus in 1781, astronomers found they had difficulty predicting its exact location at a given time. The reason, they thought, had to be the existence of another body whose gravitational pull was influencing the orbit of Uranus. In 1845 and 1846, John Couch Adams in England, and Urbain Leverrier in France, calculated the mass and location such a body would need to have. In the fall of 1846, German astronomer Johann Gottfried Galle discovered Neptune in the pre-

Earth

When *Voyager 2* passed Neptune in 1989, it had been traveling for about 12 years and had already gone nearly 3 billion miles (4.8 billion km).

dicted location. Within a few weeks the British astronomer William Lassell discovered Neptune's largest satellite, Triton. A second satellite, Nereid, was discovered more than a century later in 1949.

In August 1989 *Voyager 2* completed its tour of the gas giants, passing through the Neptunian system and coming within about 3,000 miles (4,900 km) of the planet. The wealth of data returned by the probe remains unsurpassed to date as a source of knowledge about this most distant gas giant and its moons.

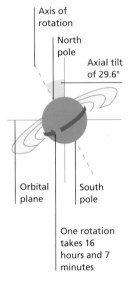

Axis of rotation

North pole

Axial tilt of 29.6°

Orbital plane

South pole

One rotation takes 16 hours and 7 minutes

Composition, Interior and Atmosphere

Neptune resembles Uranus in composition and the structure of its interior. Hydrogen, for example, contributes only about 15 percent of the planet's total mass. Water, methane, and ammonia are also in greater abundance in Neptune than in Jupiter or Saturn. Neptune's interior consists of a dense core of melted rock and ice. Because of Neptune's greater density, its core is probably slightly larger than the core of Uranus. A mantle of liquid hydrogen surrounds the core and lies beneath what may be an ocean of water mixed with methane and ammonia.

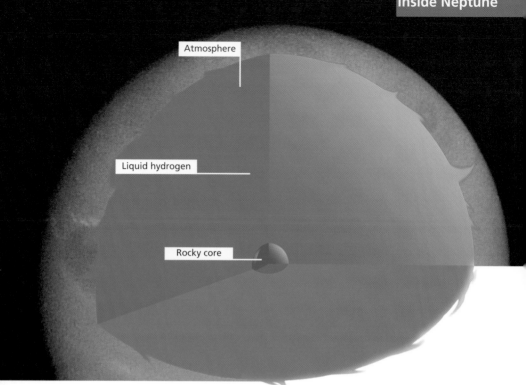

Atmosphere

Liquid hydrogen

Rocky core

Neptune's atmosphere is a mix of hydrogen, helium, and methane. Like the other gas giants, cloud layers of ammonia, ammonium hydrosulfide, and water ice are thought to exist. Temperatures in the upper atmosphere, as on Uranus, are cold enough to freeze methane. Bright clouds of methane ice form above the other cloud layers, casting shadows on the upper cloud deck. Neptune's atmosphere is much more active than Uranus's. It has zonal bands where westward moving winds reach speeds up to 1,200 miles (2,000 km) per hour. Thus Neptune has the strongest winds of all the planets. Like Jupiter and Saturn, Neptune radiates more energy than it receives from the Sun, which indicates the presence of an interior heat source. It is this source that provides the energy to drive the planet's winds as well as its large ovals, or hurricane-like storms.

Voyager 2 returned images of one of Neptune's storms—the Great Dark Spot—located at about the same latitude as the Great Red Spot on Jupiter. In 1994 images from the Hubble Space Telescope showed that the Great Dark Spot had vanished—in contrast to Jupiter's Great Red Spot, which has existed for hundreds of years.

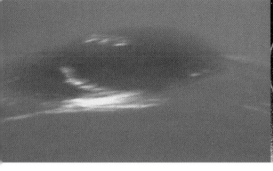

From 3.8 million miles (6.1 million km) away, the bright white clouds surrounding Neptune's Great Dark Spot are much more prominent than in the earlier, more distant image shown on p. 170.

Magnetic Field

Neptune generates a strong magnetic field that is 25 times greater than Earth's. The characteristics of Neptune's and Uranus's magnetic field are very similar. For example, Neptune's magnetic axis is tilted 47° from Neptune's rotational axis. Also, the center of the magnetic field is offset from Neptune's center by more than half the radius of the planet. Neptune's magnetic field, like that of Uranus, is probably generated where the pressure is high enough for water to conduct electrical currents rather than deep in the interior.

Rings

During an occultation of a star by Neptune in 1984, scientists observed what appeared to be an incomplete ring system consisting of three small ring arcs. *Voyager 2* images revealed that the arcs are part of one narrow ring that has three particularly dense areas. They also revealed another narrow ring and two rings that are more spread out. The ring closest to Neptune is about 26,000 miles (42,000 km) above the cloud tops. The outer ring, which contains the arcs, is about 39,000 miles (63,000 km) away.

Like the rings around Uranus, Neptune's consist of dark, non-reflective particles that range in size from microscopic dust to objects that can be measured in inches and feet. The microscopic dust is much more abundant than in the rings of Uranus. The largest ring objects are probably about 33 feet (10 m) across.

A time exposure captured this *Voyager* image (top) of Neptune's rings that are composed of small particles of dust-like material. Neptune's high-altitude clouds (bottom) resemble Earth's cirrus clouds.

Satellites

Voyager 2 images revealed six new satellites in orbit around Neptune. They range in diameter from 30 to 250 miles (50 to 400 km) and lie inside the orbit of Triton. The inner four satellites orbit within the ring system. Beyond the rings are Larissa and Proteus. After Triton, Proteus and Nereid are the largest of Neptune's moons. Nereid is the farthest away from Neptune.

Triton The most unusual of Neptune's moons, Triton is also one of the most intriguing bodies in the solar system. It has a diameter of 1,680 miles (2,700 km), about two-thirds the width of Earth's Moon. Its average density of 2.1 suggests that it is about 75 percent rock or silicates and 25 percent water ice. Of the gas giant satellites, only Io and Europa have higher average densities.

Triton's surface is among the most complex of the icy satellites. It has relatively few impact craters which indicates that icy flows from the interior caused extensive resurfacing. The equatorial region is dominated by a "canteloupe" terrain of pits or depressions crisscrossed by ridges. This terrain may have developed from multiple episodes of melting and collapse of the ice crust. Long, narrow valleys rimmed by ridges cross the surface. These appear to be cracks or fractures in the crust that filled with ice. Smooth plains rimmed by steep vertical scarps hundreds of feet high resemble volcanic calderas on Earth. The plains may have been formed when floods of icy flows created lake-like landforms following collapse of the crust. In some areas the surface is marked by dark features surrounded by bright aureoles, or rings.

Fourteen images were combined to produce this computer-enhanced image of Neptune's moon, Triton. The bright south polar cap (bottom), is marked by streaks that are believed to be dark organic compounds deposited by geysers.

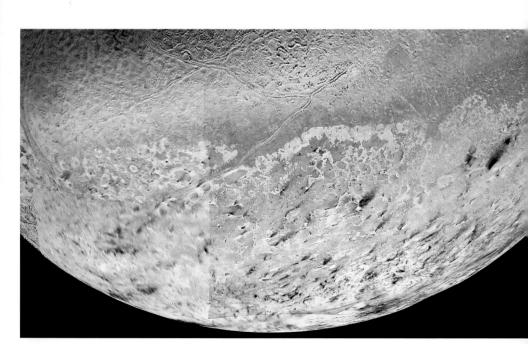

A cap of frozen methane and nitrogen covers most of the south polar region. At about −409°F (−245°C), the surface temperature there is the coldest known in the solar system. A thin atmosphere of nitrogen and methane surrounds Triton.

Evidence of geyserlike plumes were found in *Voyager 2* images. Probably caused by the explosive release of nitrogen gas, the plumes of dark material extend as high as 5 miles (8 km) above the polar caps.

Cosmic rays and ultraviolet light create new complex organic compounds from methane. These dark compounds may explain why the plumes of the geysers and the streaks they make on the surface appear black. Other organic compounds may lend the polar caps their pink tinge.

Clues to Triton's tectonic activity and extensive resurfacing may lie in its unusual orbit, which is steeply inclined to the plane of Neptune's equator, rings, and other satellites. Further, it is orbiting clockwise—opposite to the direction of the other satellites and Neptune's rotation. This may be because Triton was captured by Neptune, or a large impact may have knocked it into its current orbit. Either process would have generated enough heat to melt Triton's interior and drive its tectonics and resurfacing.

Understanding Triton may give us insight into the nature of the outermost planet of our solar system—Pluto.

A close-up of Triton's south polar caps shows several black wind streaks, or plumes, all blowing in the same direction. This points to Neptune's atmosphere as the source of the wind that blows the plume material.

Part of the complex geologic history of icy Triton (left), Neptune's largest moon, is shown in this *Voyager 2* image of August, 1989. The two depressions are thought to be the result of the collapse of the crust followed by flooding by icy flows.

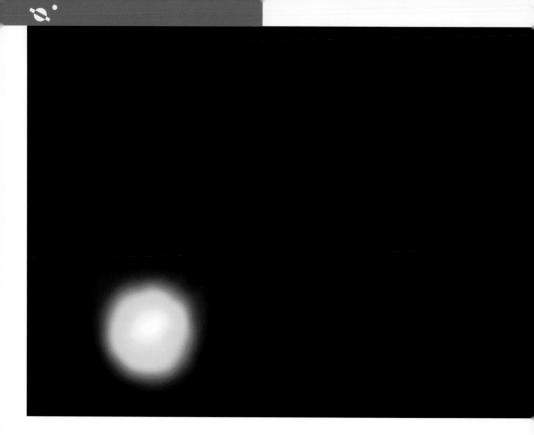

On February 21, 1994, NASA's Hubble Space Telescope took this image of Pluto and its moon, Charon. The European Space Agency's faint object camera was used to get the clearest view yet of the planet from a distance of 2.6 billion miles (4.4 billion km).

Pluto

Pluto has more in common with Triton, the largest moon of Neptune, than it does with any of the other eight planets. With a diameter of 1,440 miles (2,320 km), Pluto is actually smaller than Triton. In fact, many astronomers and planetary scientists do not classify Pluto as a planet.

Discovery

Irregularities in the orbits of Uranus and Neptune led to the suggestion that there might be a ninth planet. William Pickering and Percival Lowell, both of the United States, especially urged a systematic search for another planet, which began in 1905 and ended in 1930 with the discovery of Pluto by Clyde Tombaugh. As it turns out, however, Pluto is much too small and too distant to influence the orbits of Uranus and Neptune.

Efforts in 1978 to pinpoint Pluto's position led to the observation of a bump on its image in the photographic plate. The bump turned out to be a satellite, and it was given the name Charon.

Because Pluto and Charon are an average distance of 3.7 billion miles (6 billion km) from the Sun, Earth and space-based

In Greek mythology Pluto was the god of the underworld, and Charon was the boatman who ferried souls there.

telescopes provide limited information about them. To date, no planetary probe has investigated Pluto. But plans for a small probe called the Pluto Fast Flyby mission are under way and it could be launched as early as 1998. Twin spacecraft would reach Pluto after seven or eight years.

Pluto's Origin

The average densities of Pluto and Charon are about 2.0, the same as Triton's. Like Triton, they are probably about 75 percent rock and 25 percent water ice. One theory is that Pluto and Triton formed at the same time in the same part of the solar nebula with many other bodies of their size. Most of these bodies may have come together to form Uranus and Neptune. The gravitational influence of Neptune may have perturbed the orbits of the remaining bodies, sending some of them far beyond Pluto's orbit. Triton was captured by Neptune. Pluto may have avoided both these fates because the 3:2 ratio of its orbiting period to Neptune's protected it from the larger planet's influence.

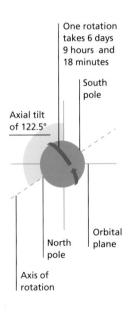

One rotation takes 6 days 9 hours and 18 minutes

South pole

Axial tilt of 122.5°

Orbital plane

North pole

Axis of rotation

These Lowell Observatory photographs recorded the discovery of the planet Pluto—the arrows indicate the slight movement that distinguished it from the surrounding mass of stars. Besides being named after the Roman god, the first two letters of its name honor American astronomer Percival Lowell.

An Extreme Orbit

Pluto's orbit is as atypical as the planet itself. With an inclination of 17.2°, Pluto rises above and drops below the ecliptic plane—the plane of the solar system—during its 248-year orbit. Pluto's orbit is so elliptical that for 20 years of its orbital period, it is closer to the Sun than Neptune, which follows a nearly circular orbit. For example, from 1979 to 1999, Pluto is traveling inside Neptune's orbit, giving Neptune the temporary distinction of being the farthest planet from the Sun.

Like Uranus, Pluto is tipped over on its side. Its rotational axis is tilted 122.5° to the plane of its orbit. As a result, there is a period of about five years during which Pluto's satellite, Charon, passes in front of and behind the planet from a vantage point of an observer on Earth. This period of *transits* (when Charon moves in front of Pluto) and *occultations* (when it passes behind) is not repeated again until the planet has gone halfway around its orbit, about 124 years later. The next series of these motions will begin in 2109 A.D.

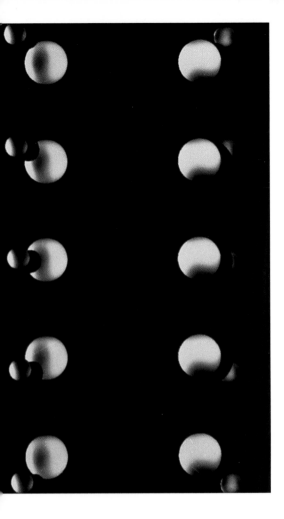

An "eclipse season" of Pluto and Charon occurs for about 6 years out of every 125 when their orbits are aligned with our view of Pluto. This model shows Charon moving southward in front of Pluto in February 1989 (left column). Three days later, Charon moves northward behind Pluto (right column).

Clues from Charon

About half the size of Pluto, Charon is 790 miles (1,270 km) in diameter, almost the same size as Saturn's moons Umbriel and Ariel. It was discovered just in time to observe a period of transits and occultations that furnished an opportunity to study the reflective properties of its surface and the surface of Pluto.

From 1985 to 1990, changes in the intensity of light reflected by Pluto and Charon were measured during each transit and occultation. A rapid decrease in brightness during a transit, for example, would indicate that Charon had covered a bright spot on Pluto. Small changes in brightness, on the other hand, might indicate that Charon was passing over a relatively darker, less reflective area. By compiling years of data and using a supercomputer, astronomers constructed maps of the bright and dark regions of Pluto and Charon. These maps indicate that polar regions on these bodies are relatively brighter than equatorial regions.

Composition and Atmosphere

Spectroscopic studies indicate the presence of methane frost on Pluto. Water frost, but no evidence of methane, has been detected on Charon. Like Triton, Pluto has a tenuous atmosphere composed of nitrogen and methane. It appears to extend out to Charon, suggesting that the two bodies share the atmosphere. Recent observations by the Hubble telescope show that Charon is bluer than Pluto. This suggests that the two bodies may have different surface compositions.

When Pluto's elliptical orbit brings it closer to the Sun, as is now the case, the planet experiences its warmest period. During this time, the nitrogen and methane frozen at the poles are released and the atmosphere becomes denser. At its closest approach to the Sun, Pluto is 30 times more distant from the Sun than Earth (30 AU). At its farthest distance from the Sun, Pluto will be 50 times more distant from the Sun than Earth (50 AU). This will occur in the year 2113. During the coldest 124 years of its orbit, all of the atmosphere will condense and fall to the surface as frost.

Collision with Neptune?

From Earth, Pluto and Neptune appear to sometimes be on a collision course. Could that ever occur?

The side view of these planets' orbits shows that at the time they might appear to collide, Pluto's orbit is well above Neptune's. Also, because the orbital periods of the planets are almost exactly in the ratio of 3:2, Pluto makes two orbits while Neptune makes three. So while Pluto makes one orbit, Neptune zooms ahead making one and a half orbits, bringing it to the opposite side of the Sun from Pluto. Thus the only time that the planets appear together from Earth is when they are, in fact, far apart and they cannot collide.

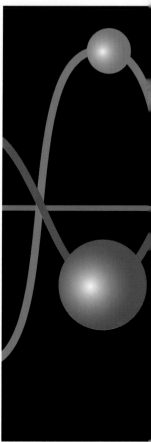

The Pluto–Charon Relationship

Charon is so large in relation to Pluto that many scientists consider the two bodies to be a double planet. One theory is that a double system was created when the two bodies came within reach of each other's gravitational fields and were mutually captured. This would help explain the wildly eccentric orbit of the Pluto-Charon pair.

A related factor supports the double planet theory. Pluto and Charon are the only planet-satellite system whose body mass ratio is very high. For most of the planet-satellite systems, the mass ratio is so low that the center of gravity—the balancing point between the two bodies—lies near the planet's center of mass. The mass ratio of Pluto and Charon is so high that the center of gravity is offset from Pluto's center of mass. The masses of the two bodies are so closely matched that the center of gravity is outside Pluto. Thus the two spiral around a common center of gravity located in the space between them.

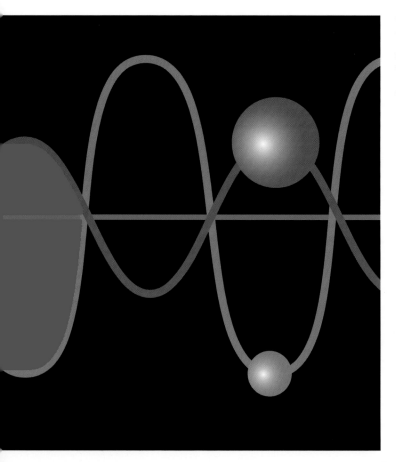

On their 248-year orbit around the Sun, Pluto (purple) and Charon (teal) spiral around a shared center of gravity, which is represented by the horizontal line.

The Farthest Reaches

Just as the universe is expanding, so is our view of the solar system. With ever-improving telescopes and probes, we continue to explore and add to our knowledge of its diverse planets and moons. Looking well beyond our solar system, we seek to further understand our place in the Milky Way galaxy as we wonder if we are unique. The big question remains: What other worlds are out there?

This spectacular view of Comet West was captured at 4 A.M. on March 7, 1976 in Sullivan, New Hampshire.

This color-enhanced image of Halley's Comet was photographed at the Lowell Observatory on May 19, 1910. The view was captured when the comet was at a distance of 84 million miles (135 million km) from the Sun and 28 million miles (45 million km) from Earth.

Comets

In addition to the nine planets and their orbiting satellites, the solar system includes a large population of small bodies called *planetesimals,* or minor planets. These minor planets are the remains and fragments of objects that did not become part of any of the major planets and satellites during their formation. They consist of two types of objects—asteroids and comets. Most asteroids are found in the asteroid belt, revolving around the Sun in orbits between those of Mars and Jupiter. Comets originate beyond the orbit of Neptune and Pluto in a region called the Oort Cloud, which may contain as many as one trillion comets.

The Kuiper Belt

Short-period comets—those with orbital periods of less than 200 years, like the famous Halley's Comet, originate from the Kuiper Belt—the innermost region of the Oort Cloud. Named after American astronomer, Gerard Kuiper, who proposed its existence in 1951, the Kuiper Belt is thought to contain millions of icy bodies. They revolve around the Sun in the same direction as the planets within 30° of the orbital plane of the planets.

The Oort Comet Cloud

In 1951 Jan Hendrik Oort, a Dutch astronomer, proposed the existence of a spherical cloud around the Sun in which most comets originate. The Oort Cloud is thought to have a radius of 20,000 to 100,000 AU, and may extend halfway to the nearest star—Alpha Centauri—which is about 275,000 AU or 4 light-years from the Sun. It has been estimated that the more than one trillion comets contained in the Oort Cloud have a total mass of only about 25 times the mass of Earth. Although the Oort Cloud has never been observed, the objects within it are believed to orbit the Sun very slowly. The gravitational force of a passing star

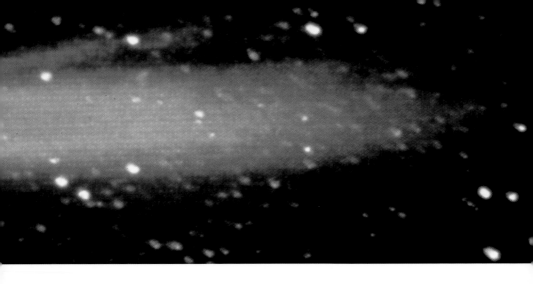

or other object can change the orbit of a comet, pulling it within the cloud into a new orbit closer to the Sun. Such comets may rotate at any inclination and may not orbit in the same direction as the planets. These comets often have orbital periods of 100,000 years or more and are called long-period comets.

What Is a Comet?

In 1950 Fred Whipple, an American astronomer, proposed that comets are composed of a mixture of ice, dust, and rock. He described them as "dirty snowballs." Further studies have confirmed Whipple's description. The nucleus of a comet appears to be largely a mixture of ice and dust, possibly with a rocky core at its center. This nucleus stays inactive until it approaches the Sun. Sunlight causes the ice to *sublimate,* or change directly from a solid to a gaseous state. When a comet comes within about three times the distance of Earth from the Sun, the increased sublimation creates the *coma*—the envelope of gas and dust surrounding the nucleus. Gas and dust from the coma are always carried away from the Sun by a combination of solar wind and sunlight. Large hydrogen clouds extending for several million miles also accompany many comets.

This black and white image of Halley's Comet was taken in March, 1986. For more than two thousand years, stargazers have been fascinated by Halley's.

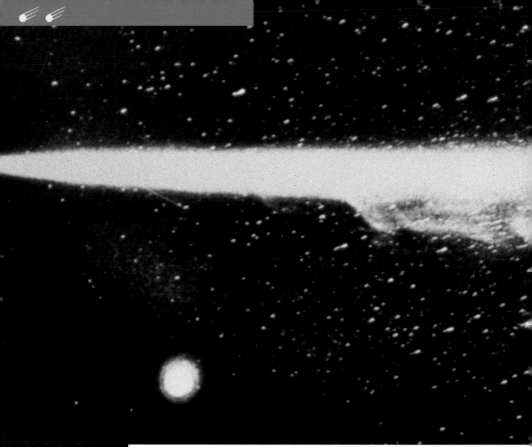

This celebrated Lowell Observatory image of Halley's Comet was taken on May 13, 1910. The planet at lower left is Venus, and the stripes at far left are the lights of Flagstaff, Arizona.

Comet Tails Comets often display two distinct types of tails, depending on the composition of the comet's particles. One type of tail is composed largely of dust particles and appears yellow or white from reflected sunlight. The dust particles within this tail are small (about 1 micron across) and are swept off the nucleus into arcs that may extend for millions of miles. The other type of tail is composed of ions (particularly carbon monoxide ions) and electrons. This tail is often straight, blue in color, and extends about ten times farther than a dust tail. The fine structures associated with these tails reflect the interaction of the ions with magnetic field lines carried in the solar wind. A comet may have either kind of tail, or both kinds at the same time.

On its first pass through the inner solar system a comet is bright. During its approach to the Sun, the nucleus loses some of its ice and dust through sublimation. At the same time, a dust layer may accumulate on the comet, becoming thicker with each pass. For this reason short-period comets, which pass close to the Sun relatively often, are generally dim. Long-period comets, which seldom get close enough to the Sun to lose much of their icy content, remain bright.

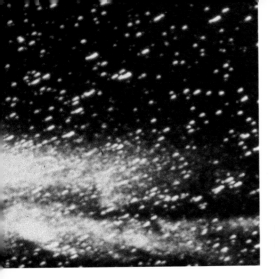

A comet's tail may be millions of miles long, but all the matter in it would fit into a suitcase.

Halley's Comet

In 1705 Edmund Halley used Newton's newly developed theories to analyze the orbits of bright comets that had been observed in 1531, 1607, and 1682. He concluded that all three were the same comet, with a period of 75 or 76 years. He predicted that this comet would return in 1758. It did, and although Halley had died 17 years earlier, the comet was named in his honor. We now know that there have been a total of 30 recorded sightings of Halley's Comet since 240 B.C. Much of our knowledge about comets is based on these observations, especially those made when it returned to the inner solar system in 1986.

Probes to Halley's Comet

In 1986 five spacecraft specifically designed to study Halley's Comet were already in space and poised for their encounters with the comet. Two of these missions were launched by the Soviets, two by Japan, and one by the European Space Agency (ESA). The United States decided against a dedicated, special mission to study Halley's Comet and used its existing deep-space satellites instead.

Studying Halley's Comet from space was difficult particularly because the comet moves in a *retrograde* orbit—that is, it moves in the opposite direction from the planets. Any spacecraft launched from Earth moves in the same direction as Earth. Because the comet would pass the spacecraft going in the opposite direction, analysis of Halley's Comet would have to be done very quickly. The spacecraft would approach the comet at about 157,000 miles per hour (70 km per second), passing through the coma—approximately the distance from Earth to the Moon—in about an hour and a half. Danger from impacts by high-velocity particles would be great and imaging difficult.

Giotto, the Artist

The European Space Agency (ESA) named its *Giotto* probe after an Italian painter who was apparently inspired by Halley's Comet nearly 700 years ago. Giotto di Bondone began painting a cycle of frescoes inside the Scrovegni Chapel in Padua around 1303. Two years earlier, in 1301, Halley's Comet had made one of its regular appearances. In Giotto's fresco called "Adoration of the Magi," the Star of Bethlehem appears as a blazing red comet in the sky above Christ's birthplace.

Vega 1 and 2 Launched in 1984, the Soviet missions *Vega 1* and *Vega 2* were designed to study first the planet Venus and then Halley's Comet. On March 6 and 9, 1986, they became the first probes to fly past Halley's at a distance of 4,990 to 5,525 miles (8,030–8,890 km). *Vega 1* traveled close enough to provide measurements of Halley's nucleus and images with as good as 660 feet (200 m) resolution. In all, it transmitted 500 images and returned data on the structure and chemistry of the comet. *Vega 2* collected more images and data and viewed the side of Halley's elongated nucleus.

Susei and Sakigake Japan's spacecraft *Susei* and *Sakigake* were to study the comet's *corona*—its extensive hydrogen cloud—and monitor the solar wind. Although they viewed Halley's from farther away than the *Vegas*, *Susei* was hit by dust particles from the comet even at a distance of 93,800 miles (151,000 km). On March 8, between the *Vega* encounters, *Susei* observed that the brightness of the hydrogen cloud surrounding Halley's varied on a cycle of about 53 hours, the estimated rotation period for the comet's nucleus. On March 11 *Sakigake* gathered data on the solar wind around the comet. Both spacecraft are currently en route to encounters with other comets, including Giacobini-Zinner (November 1998) and Temple-Tuttle (February 1998).

Giotto The ESA probe *Giotto* approached Halley's less than a week after *Vega 2*. It passed within 376 miles (605 km) of the comet's nucleus—closer than any other probe. To survive the impact of dust particles, *Giotto* had a two-layer shield. Despite being battered by a hail of high-velocity dust particles, *Giotto* returned the highest resolution images of Halley's inner coma and its nucleus.

U.S. Missions The U.S. spacecraft that collected data in the vicinity of Halley's Comet were *Pioneer Venus*, *Pioneer 7*, and *Dynamics Explorer 1*. In addition, the International Sun-Earth Explorer (ISEE 3) satellite was also redirected to investigate Halley's. It was renamed the International Cometary Explorer (ICE) to designate the new focus of its mission.

Probe Results

The probes revealed significant data about Halley's nucleus. It is irregularly shaped, resembling a potato about 10 miles (16 km) long and about 5 miles (8 km) across at its widest. The surface of the nucleus is very dark—darker than coal—comparable to the darkest objects in the solar system. The nucleus also has surface features such as hills and what are believed to be craters. Bright jets of sublimated gas-carrying dust particles become active at sunrise and inactive at sunset. This suggests that a dust-rich crust may be thin in the area of the jets, allowing the outgassing. The comet's inner coma consists of a mixture of about 80 percent water vapor, 10 percent carbon monoxide, and 3.5 percent carbon dioxide, and some complex organic compounds. Most of the particles in the tail are composed of either a mixture of hydrogen, carbon, nitrogen, and oxygen or silicates.

Close-up shots of Halley's Comet were obtained by ESA's spacecraft, *Giotto* (left) in March, 1986. The spacecraft flew within 376 miles (605 km) of the comet's icy nucleus (right) which proved to be darker and more dust-covered than astronomers expected.

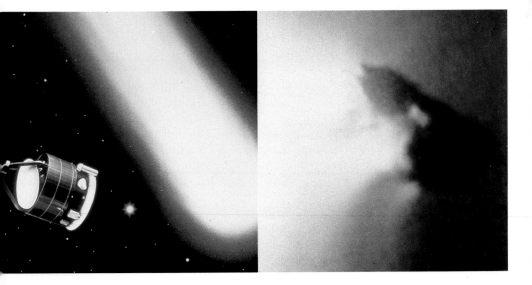

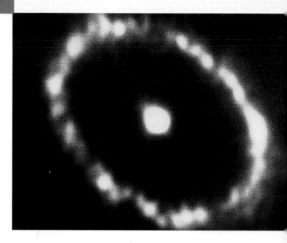

"A day will come when beings shall stand upon this Earth, as one stands on a footstool, and laugh and reach out their hands amidst the stars."
—H.G. Wells, 1902

Hubble Space Telescope returned this image of a ring around Supernova 1987A. The ring is an envelope of gas ejected after the massive supergiant star exploded.

Planet X

After the discovery of Pluto (and the realization that Pluto was too small to cause the deviation observed in the orbits of Uranus and Neptune) there was a great deal of speculation that other, more distant planets might exist. Pluto's discoverer, Clyde Tombaugh, searched the skies, comparing photographic plates for thirteen years before concluding that there were no additional planets. However, it was noted that Halley's Comet, observed for hundreds of years, consistently arrived several days later than the computed time. Something as yet undiscovered seemed to be disturbing its orbit as well. Once again there was much debate regarding a possible Planet X beyond Pluto.

Modern observations of Neptune, Uranus, and Halley's Comet indicate that there are no deviations in their orbits. However, this may not have been the case when they were first noted in the 1800s. To account for this difference, some astronomers theorize that a large planet does indeed exist beyond Pluto, but that its highly elliptical orbit is steeply inclined to the orbital plane of the other planets. They believe that in the 1800s it was close enough to affect the orbits of Uranus, Neptune, and Halley's Comet, but is now too far away to influence these bodies. Some have speculated that this undiscovered planet has five times the mass of Earth, orbits the Sun once in 1,000 years, and moves in an orbit perpendicular to that of Earth.

Another explanation is that there is not one Planet X but several. Some astronomers suspect that several mini-planets beyond Pluto may cause the observed deviations. Indeed, six objects were discovered orbiting the Sun beyond Pluto in 1992 and 1993. These bodies, probably members of the population of objects within the Kuiper Belt, have diameters of about 150 miles (240 km) and orbit about 50 times farther from the Sun than Earth. It is possible that such objects in the inner part of the Kuiper Belt may influence the orbits of larger bodies at times and account for observed deviations.

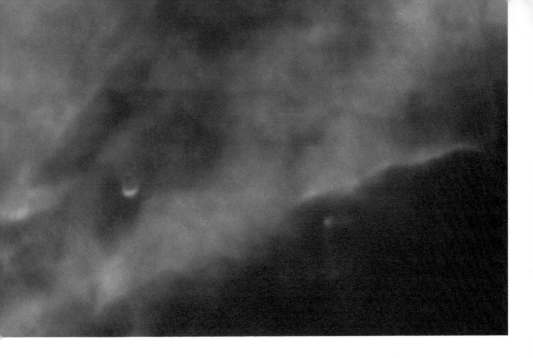

Searching the Stars for Other Planets

In a galaxy that contains billions of stars, it would be hard to imagine that our Sun is the only star orbited by planets. In fact, since our Sun is a rather typical star in the Milky Way galaxy, one might expect that planetary systems would be the rule rather than the exception. Although even the closest neighboring stars are too distant to separate the reflected light from a planet from the light of its sun, there are other ways we can search for solar systems. One method involves careful measurements of the changes in the intensity of light coming from a star. If the star is orbited by a planet or planets, the intensity of the star's light will drop slightly as a planet passes between the star and the observer on Earth. If such periodic variations can be detected, it would be strong evidence of the existence of a planetary system.

Another method involves the very accurate determination of a star's motion. A star orbited by a planet will move about the system's center of gravity. If the planet is massive enough, the star's motion about a common center of mass will cause it to have a distinct wobble. In 1963, Barnard's Star became the first example of a possible planetary system detected in this way.

A wobble of a star can also be detected by a shift in the wavelength of the star's light as it moves toward or away from an observer on Earth. This is called the *Doppler effect*. The same effect causes the change in pitch of an automobile horn as the car approaches and then passes you. With spaceborne telescopes, it may soon be possible to detect Doppler shifts caused by Jupiter-sized planets orbiting sun-like stars.

This close look at a 1,500-year-old gas plume in Orion nebula was captured by NASA's Hubble Space Telescope.

PSR1257+12 is so dense that a teaspoonful of its matter would weigh more than a trillion tons.

PSR1257+12: Another Solar System

As it turns out, the best candidate for another planetary system in our galaxy involves a *pulsar*, rather than a normal star. A pulsar is the ultra-dense remnant of a supernova—a massive star that exploded. Rotating at incredible speeds, pulsars emit radio waves generated by charged particles trapped in their powerful magnetic fields. The period of the radio pulses that sweep across Earth from pulsars is extremely regular. It is this property that led to a momentous discovery.

The Doppler Effect and Light

The Doppler effect is commonly observed as it applies to sound. As the diagram below illustrates, the wavelength of sound waves are reduced as the car moves toward the listener and increased as the car moves away—causing the familiar rise and fall in pitch. The same principle holds true for light waves. Red light has the lowest frequency, or longest wavelength, in the visible spectrum. Violet light, on the other end of the visible spectrum, has the highest frequency, or shortest wavelength. Therefore, light sources moving toward the observer have spectrums shifted toward the violet end, while those moving away shift toward the red.

Using a radio telescope at the Aricebo Observatory in Puerto Rico, astronomer Alexander Wolszczan observed irregularities in the period or flash rate of radio waves from a pulsar in the constellation Virgo. This pulsar, some 1,300 light-years from Earth, was designated PSR1257+12. After collecting data for several years, Wolszczan concluded that the only explanation for the variations was a reflex motion of the pulsar caused by the presence of three orbiting planets. In 1994 Wolszczan announced that a new solar system had been "unambiguously identified."

Two of the planets in this system, designated Planets A and B are about three times the mass of Earth. The third planet is slightly more massive than Earth's Moon. Planet A is the closest to the pulsar, orbiting it in 25 days at a distance of only 0.19 AU, about half the distance of Mercury from the Sun. Planet B orbits the pulsar in 67 days at a distance of 0.36 AU, a little closer than Mercury. Planet C is the farthest from the pulsar, orbiting it in 98 days at a distance of 0.47 AU, about half the distance between the orbits of Mercury and Venus.

Because these planets are associated with a pulsar, it is not clear whether they are the remains of a planetary system destroyed by the supernova or were formed from material left in its wake. In any event, it is unlikely that life as we know it exists on the planets of PSR1257+12. The pulsar is about 4.7 times more radiant than the Sun and the planets are probably enveloped in a dense cloud of charged particles.

This is only the first of many planetary systems that will probably be discovered in the years to come. Some of these will undoubtedly be solar systems similar to our own.

Alexander Wolszczan (left) poses near computer-generated images of his discovery—pulsar PSR1257+12. The Arecibo Observatory in Puerto Rico (right), from where the pulsar discovery was made, is operated by Cornell University and funded by the National Science Foundation.

Comparing the Planets

The planets and satellites of our solar system are remarkably diverse. The four inner, Earth-like planets are composed of rocky minerals and metals and only two have satellites. The four outer gas giants are composed of mainly hydrogen and helium. All have ring systems as well as satellites, and the satellites are even more diverse. However, the planets and satellites also have much in common. Here they are compared to illustrate some of their differences and similarities.

A collage of Saturn and five of its moons provides a glimpse of the great differences between bodies in the solar system.

Our Solar System

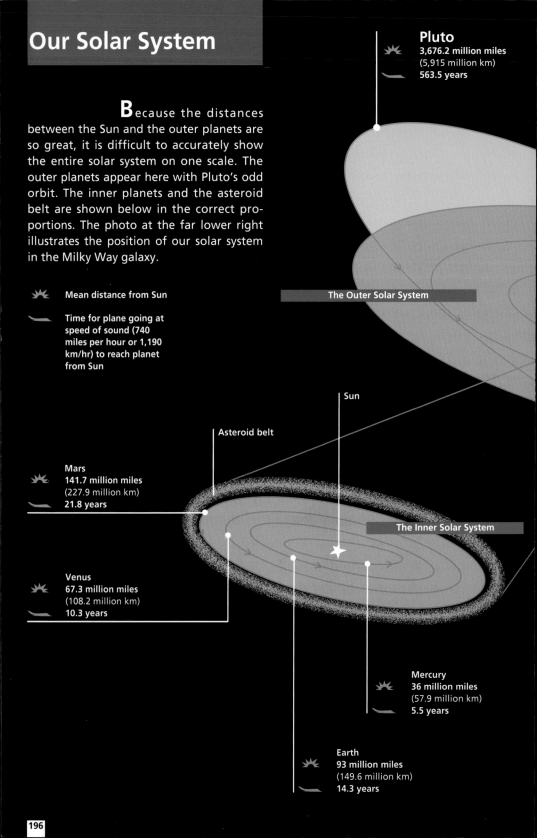

Pluto
3,676.2 million miles
(5,915 million km)
563.5 years

Because the distances between the Sun and the outer planets are so great, it is difficult to accurately show the entire solar system on one scale. The outer planets appear here with Pluto's odd orbit. The inner planets and the asteroid belt are shown below in the correct proportions. The photo at the far lower right illustrates the position of our solar system in the Milky Way galaxy.

Mean distance from Sun

Time for plane going at speed of sound (740 miles per hour or 1,190 km/hr) to reach planet from Sun

The Outer Solar System

Sun

Asteroid belt

Mars
141.7 million miles
(227.9 million km)
21.8 years

The Inner Solar System

Venus
67.3 million miles
(108.2 million km)
10.3 years

Mercury
36 million miles
(57.9 million km)
5.5 years

Earth
93 million miles
(149.6 million km)
14.3 years

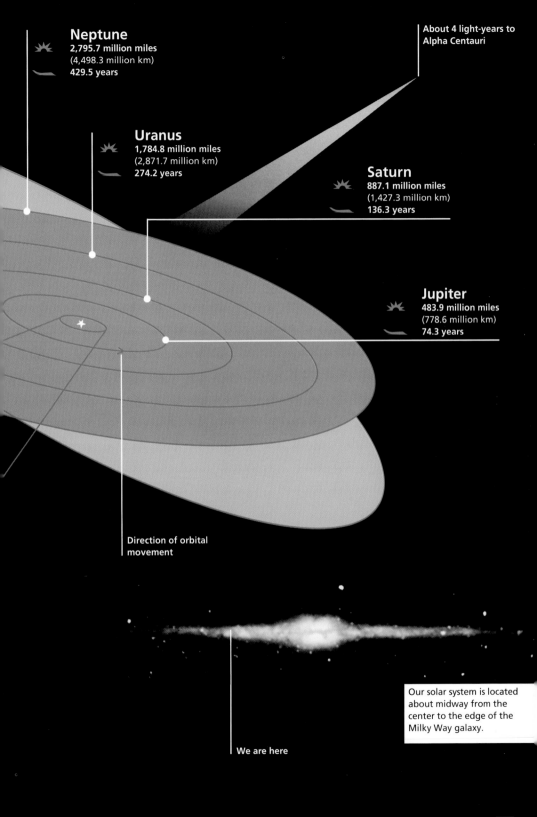

Neptune
2,795.7 million miles
(4,498.3 million km)
429.5 years

About 4 light-years to Alpha Centauri

Uranus
1,784.8 million miles
(2,871.7 million km)
274.2 years

Saturn
887.1 million miles
(1,427.3 million km)
136.3 years

Jupiter
483.9 million miles
(778.6 million km)
74.3 years

Direction of orbital movement

We are here

Our solar system is located about midway from the center to the edge of the Milky Way galaxy.

Planets

The planets' sizes are shown here with their angle of axis and direction of rotation. All rotate in the same direction, although Venus, which is almost upside down, appears to rotate backward.

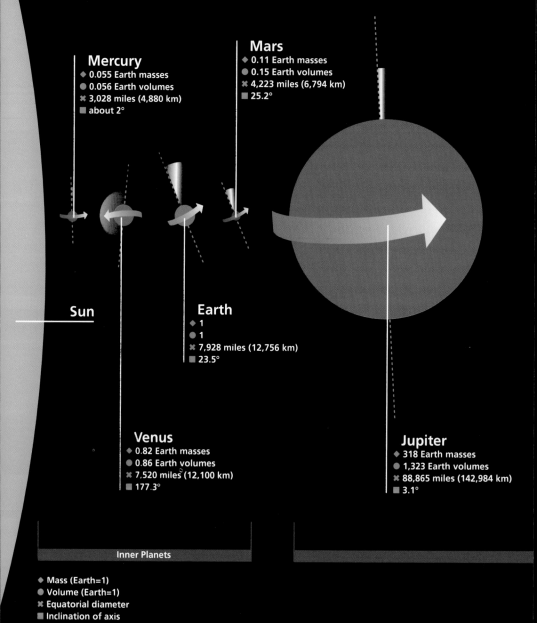

Mercury
- ◆ 0.055 Earth masses
- ● 0.056 Earth volumes
- ✷ 3,028 miles (4,880 km)
- ■ about 2°

Mars
- ◆ 0.11 Earth masses
- ● 0.15 Earth volumes
- ✷ 4,223 miles (6,794 km)
- ■ 25.2°

Sun

Earth
- ◆ 1
- ● 1
- ✷ 7,928 miles (12,756 km)
- ■ 23.5°

Venus
- ◆ 0.82 Earth masses
- ● 0.86 Earth volumes
- ✷ 7,520 miles (12,100 km)
- ■ 177.3°

Jupiter
- ◆ 318 Earth masses
- ● 1,323 Earth volumes
- ✷ 88,865 miles (142,984 km)
- ■ 3.1°

Inner Planets

- ◆ Mass (Earth=1)
- ● Volume (Earth=1)
- ✷ Equatorial diameter
- ■ Inclination of axis

● 67 Earth volumes
✶ 31,770 miles (51,118 km)
■ 97.9°

Saturn
◆ 95.18 Earth masses
● 744 Earth volumes
✶ 74,914 miles (120,536 km)
■ 26.7°

Pluto
◆ 0.002 Earth masses
● 0.006 Earth volumes
✶ 1,430 miles (2,300 km)
■ 122.5°

Neptune
◆ 17.14 Earth masses
● 57 Earth volumes
✶ 30,782 miles (49,528 km)
■ 29.6°

Outer Planets

Planets' Composition

The inner and outer planets have different compositions. The inner planets are composed mostly of silicates, the minerals that make up rocks and metals. The outer planets, with the exception of Pluto, are gaseous. Pluto is so small in relation to its neighbors, it would be almost invisible on this chart.

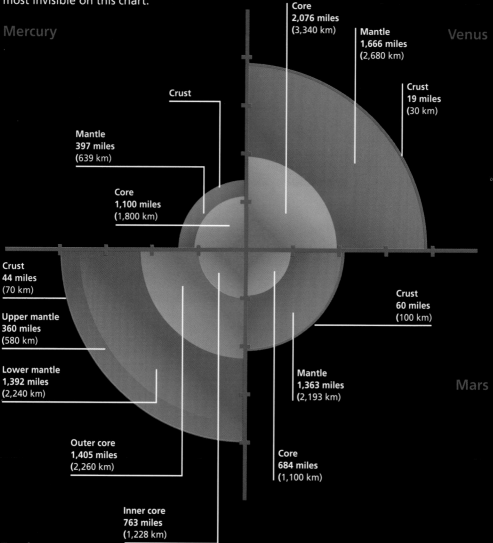

Mercury

Venus

Core
2,076 miles
(3,340 km)

Mantle
1,666 miles
(2,680 km)

Crust

Crust
19 miles
(30 km)

Mantle
397 miles
(639 km)

Core
1,100 miles
(1,800 km)

Crust
44 miles
(70 km)

Crust
60 miles
(100 km)

Upper mantle
360 miles
(580 km)

Lower mantle
1,392 miles
(2,240 km)

Mantle
1,363 miles
(2,193 km)

Mars

Outer core
1,405 miles
(2,260 km)

Core
684 miles
(1,100 km)

Inner core
763 miles
(1,228 km)

Earth

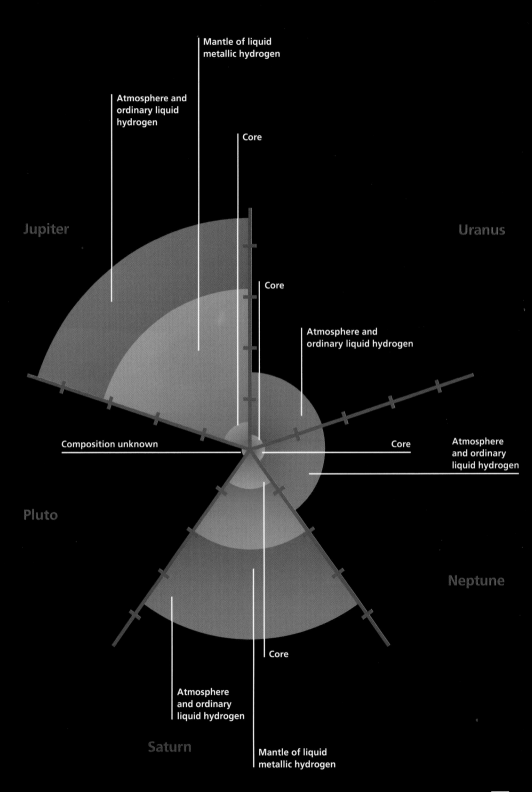

Mantle of liquid
metallic hydrogen

Atmosphere and
ordinary liquid
hydrogen

Core

Jupiter

Uranus

Core

Atmosphere and
ordinary liquid hydrogen

Composition unknown

Core

Atmosphere
and ordinary
liquid hydrogen

Pluto

Neptune

Atmosphere
and ordinary
liquid hydrogen

Core

Saturn

Mantle of liquid
metallic hydrogen

Moons

The moons of the solar system vary greatly both in their surface characteristics and their composition. Shown here with Earth's Moon are Jupiter's moons Ganymede and Io, Saturn's Titan, Uranus's Titania, and Neptune's Triton.

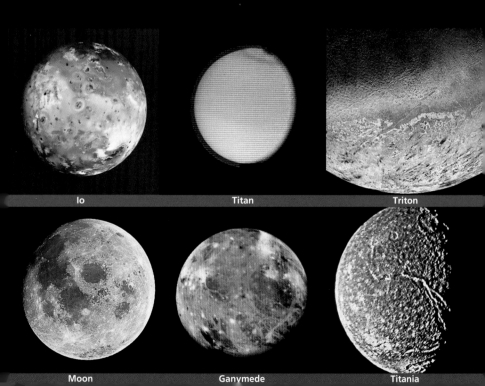

Io

Titan

Triton

Moon

Ganymede

Titania

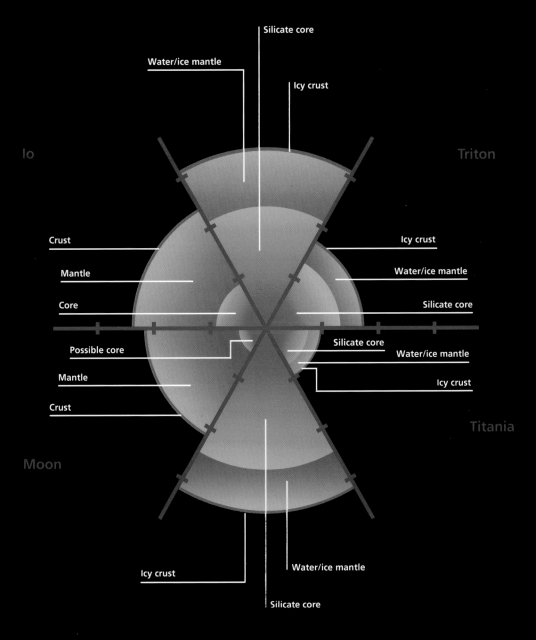

Titan

Silicate core

Water/ice mantle

Icy crust

Io

Triton

Crust

Icy crust

Mantle

Water/ice mantle

Core

Silicate core

Possible core

Silicate core

Water/ice mantle

Mantle

Icy crust

Crust

Titania

Moon

Icy crust

Water/ice mantle

Silicate core

Ganymede

Rings

Up until the late 1970s, Saturn was thought to be the only planet with rings. We now know that Jupiter, Uranus, and Neptune also have ring systems. We are only beginning to understand their complex structures.

Uranus's ring system with dust particles

Jupiter's faint ring system

Neptune's two main rings

Saturn's rings and their shadows

Views from the Surface

Much of our exploration of the planets and their satellites has been accomplished using probes and landers that imaged the surfaces of these bodies from space. Venus, the Moon, and Mars have the distinction of being viewed from their surface by robotic landers. The Moon is the only extraterrestrial body whose surface has been explored directly by humans.

Apollo 17 site, Moon

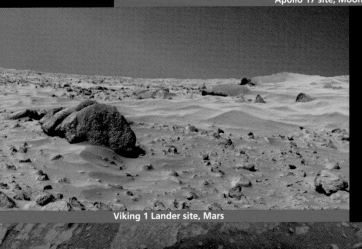

Viking 1 Lander site, Mars

Venera 13 Lander site, Venus

Craters

The solid surfaces of most of the bodies of our solar system have one thing in common—craters. Impact craters vary greatly in size and complexity. They also reflect aspects of the material in which they are formed. For example, on Mars craters with lobe-shaped, layered ejecta blankets may indicate the presence of permafrost or groundwater.

Venus

Meteor Crater, Earth

Mars

Moon

Volcanoes

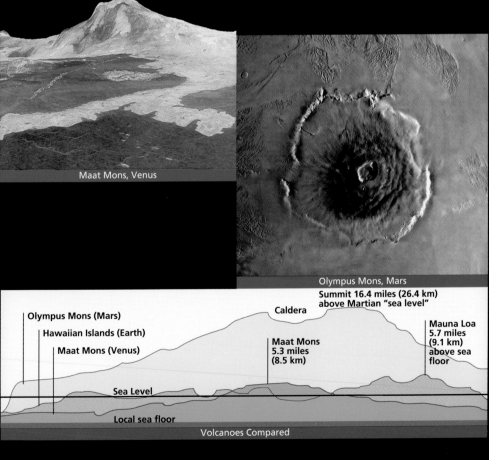

All the terrestrial planets, with the exception of Mercury, have shield volcanoes. The largest volcanoes on Venus, Earth, and Mars are broad, dome-shaped features with gentle slopes, though their sizes vary widely. Olympus Mons on Mars towers above both Venus's Maat Mons and Mauna Loa on Earth.

Mauna Loa, Hawaii

Maat Mons, Venus

Olympus Mons, Mars

Summit 16.4 miles (26.4 km) above Martian "sea level"

Olympus Mons (Mars)

Hawaiian Islands (Earth)

Maat Mons (Venus)

Caldera

Maat Mons 5.3 miles (8.5 km)

Mauna Loa 5.7 miles (9.1 km) above sea floor

Sea Level

Local sea floor

Volcanoes Compared

Discovery Rupes, Mercury

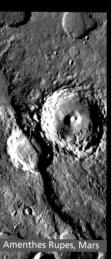

Amenthes Rupes, Mars

The influence of tectonic forces is evident on nearly every major body in the solar system that has a solid surface. Forces that compress planetary crusts create fault scarps like Discovery Rupes on Mercury and Amenthes Rupes on Mars, and wrinkle ridges like those in the volcanic plains on Mercury, Earth, and Mars. Tectonic forces that stretch planetary crusts form fracture and rift systems like Claritas Fossae on Mars and Devana Chasma on Venus.

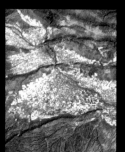

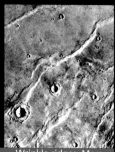

Wrinkle ridges, Mercury

Wrinkle ridges, Earth

Wrinkle ridges, Mars

Claritas Fossae, Mars

Devana Chasma, Venus

Adams, John Couch
(1819–1892) An English mathematician and astronomer who, during the years 1843–1845, deduced the existence and location of Neptune by mathematical calculation. *See* **Galle, Johann.**

advection The horizontal movement of a mass of gases or fluids, which provides a means for the transfer of heat.

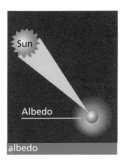
albedo

albedo The ratio of the light reflected by an object, such as a planet, to the total light that falls upon it.

Amalthea The fifth-largest satellite of Jupiter, discovered by telescope in 1892.

annular eclipse An eclipse of the Sun in which a thin ring of the Sun's outer disk remains visible around the dark disk of the Moon.

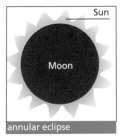
annular eclipse

aphelion The point farthest from the Sun in the orbit of a planet, comet, or asteroid. *See* **perihelion.**

Aphrodite Terra The second-largest highland area on Venus. It is about the size of the United States and extends nearly halfway around the planet in the equatorial region.

apogee The point farthest from Earth in the orbit of the Moon or of an artificial satellite that is in orbit around Earth. *See* **perigee.**

Apollo Program NASA's lunar exploration program that began in 1967 and ended in 1972. Astronauts of the *Apollo 11* mission of July 1969 became the first humans to walk on the surface of the Moon.

arachnoids Surface structures on Venus that resemble the filaments of a spider's web and are thought to have been created by molten magma rising up from the interior and cracking the planet's crust.

Argyre Basin

Argyre An impact basin on Mars that is 750 miles (1200 km) across and nearly 2 miles (3.2 km) deep.

Ariel

Ariel A satellite of Uranus that is about 721 miles (1,160 km) in diameter and is characterized by light and dark patches, cratering, and resurfacing.

asteroid Any of the rocky celestial bodies, measuring from a few hundred feet to several

Gaspra asteroid

and the Sun: 93,000,000 miles (150,000,000 km).

Atlas A very small satellite of Saturn that appears to serve as a shepherd satellite in Saturn's ring system.

atmosphere A gaseous envelope surrounding some planets and moons. Gravity holds an atmosphere close to the surface of the body it surrounds.

atmospheric pressure The pressure exerted by

a planet's or satellite's atmosphere at any given point, usually calculated in terms of the normal pressure of Earth's atmosphere at sea level, or about 14.5 pounds per square inch (1 kg/cm2).

AU *See* **astronomical unit.**

aurora borealis A luminous phenomenon in the upper atmosphere of Earth's northern hemisphere. It is caused by the emission of light from air molecules being excited by charged par-

hundred miles in diameter, the majority of which are found in the asteroid belt. Also called *minor planet*.

astronomical unit (AU) A unit of length in astronomy that is equal to the average distance between Earth

Asteroid Belt

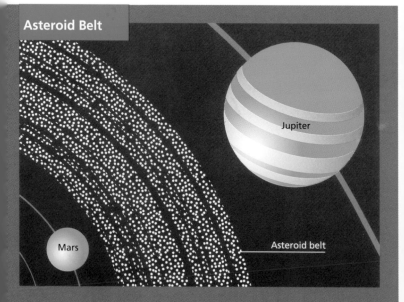

Jupiter

Mars

Asteroid belt

Most asteroids orbit around the Sun in the interplanetary region between Mars and Jupiter. Following somewhat elliptical orbits in the same direction as the planets, the asteroids take from three to six years to complete an orbit. Asteroids are debris left over from the formation of the Solar System.

ticles from the solar wind. These particles are trapped by Earth's magnetic field. Also called *northern lights.*

autumnal equinox *See* **equinox.**

basalt A dark, fine-grained rock formed by the solidification of lava and composed mainly of silicate minerals.

Beta Regio A highland region on Venus that is in the northern hemisphere and consists of a pair of volcanic peaks.

Brahe, Tycho (1546–1601) A Danish astronomer who amassed a very accurate set of observations on the positions of stars and planets.

caldera A large volcanic crater or depression. It is usually the result of collapse following the withdrawal of underlying magma.

Callisto The outermost of Jupiter's Galilean satellites. It has a diameter of 2,985 miles (4,800 km), about the same as that of Mercury, and is one of the most heavily cratered objects known among the solid-surface planets and satellites.

Caloris Basin An impact basin approximately 810 miles (1,300 km) across on the planet Mercury.

canal

canal Any of the dark lines that were once thought to crisscross the surface of the planet Mars. In 1877 the Italian astronomer Giovanni Schiaparelli reported seeing these lines, calling them by the Italian word *canali.* Some later astronomers, including Percival Lowell, took the word to mean "canals." This led to the idea that irrigation canals had been built by intelligent life and that Mars was, or had been, inhabited. In 1971 observations by

the Mariner 9 space probe proved that there are no canals on Mars.

Cassini Division or **Cassini's Division** The apparent gap between the two brightest rings of the planet Saturn. The Division contains far fewer particles of matter than the rings. It was named for its discoverer, Jean-Dominique Cassini.

Cassini, Jean-Dominique (Giovanni Domenico) (1625–1712) An Italian astronomer who was the first director of the Paris observatory.

Cassini mission A mission, planned by NASA and ESA for 1997, to explore the planet Saturn and its system of rings and moons. An orbiter will return images and other data and a probe will make a descent through the satellite Titan's atmosphere.

celestial body Any of the various objects occupying the universe, such as galaxies, stars, planets, moons, and asteroids.

celestial equator A great circle on the celestial sphere. It is defined by the plane of Earth's equator.

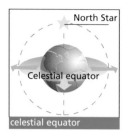
celestial equator

celestial sphere An imaginary sphere of infinite extent seeming to surround Earth and against which the celestial bodies appear to be located as viewed from Earth.

Charon The satellite of the planet Pluto, discovered in 1978. At about half Pluto's diameter, Charon is the largest satellite in the solar system in proportion to its parent planet.

chromosphere The region of the Sun's atmosphere that is between the photosphere and the corona and is composed of a thin layer of gases only a few thousand miles thick.

Clementine A robotic orbiter launched on January 25, 1994. It surveyed the Moon for several months, returning images and data on surface features. A de-

tailed, global mosaic of the surface was made using the images.

Cleopatra A large impact crater near the summit of Maxwell Montes on the planet Venus.

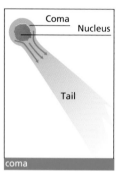
coma

coma A diffuse envelope of dust and gas that often surrounds the nucleus of a comet.

comet. See feature box on page 213.

conduction A mechanism of heat transfer through matter by the transference of heat energy from one particle to another without a net displacement of the particles.

continental drift The slow lateral movement of Earth's continents caused by the motion of the lithospheric plates.

convection A mechanism of heat transfer through a fluid or gas.

As the fluid is heated, it expands and becomes lighter than the surrounding cooler material. The hot material rises while the colder surface material sinks.

co-orbital satellites *See* page 214.

Copernicus, Nicolaus (1473-1543) A Polish astronomer who proposed that Earth rotates on its axis once each day and revolves, along with the other planets, around the Sun.

core The dense central part of a celestial body, such as a planet or the Sun.

corona¹ Any of several circular or elliptical features formed on the surface of the planet Venus. They appear to be the result of volcanic or tectonic activity.

corona² The tenuous outer layer of the Sun's atmosphere, extending millions of miles into interplanetary space.

cosmic rays Very high-energy, charged particles of unknown origin that continually enter the solar system from outer space at close to the speed of light.

Comet

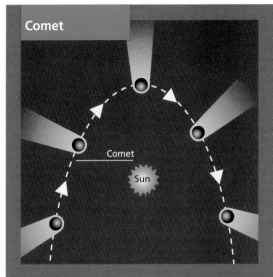

Comet

Sun

A celestial body consisting of a solid nucleus made of ice and dust often surrounded by a coma. Many comets make long, elliptical orbits around the Sun. In the part of its orbit closest to the Sun, a comet develops a bright halo and a long vaporous tail that frequently increase the comet's visibility from Earth. The tail points away from the Sun and can be 100 million miles (161 million km) long. Each time a comet passes close to the Sun, it loses part of its ice and eventually becomes one of the group of asteroids that approach Earth.

crater A bowl-shaped depression on the surface of a terrestrial planet or a moon, caused by the extrusion of volcanic material. See **impact crater**.

crater chain A series of craters on the surface of a planet or moon. A crater chain is formed by the fragments of an asteroid or comet that broke apart before impact.

crust The solid outer layer of a planet or moon. It is composed of materials having relatively low density and low melting points.

D

day The period of time that a planet or moon takes to make one complete rotation on its axis. For example, a day on Uranus is 17 hours; on Venus a day is equal to 243 Earth days.

density A measurement of how much mass is contained in a given volume of space. Water has a density of 1.0, or one gram per cubic centimeter. By this measurement, Earth's average density is 5.52.

Dione A satellite of Saturn discovered by Jean-Dominique Cassini in 1684. It has a diameter of nearly 700 miles (1,120 km).

E

eclipse

eclipse An alignment of three celestial bodies such that the middle body partially or totally

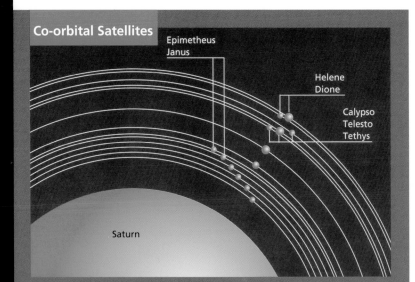

Co-orbital Satellites

Epimetheus
Janus

Helene
Dione

Calypso
Telesto
Tethys

Saturn

Satellites that share very nearly the same orbit. The planet Saturn has three groups of co-orbitals: Janus and Epimetheus; Tethys, Telesto and Calypso; and Dione and Helene. About every four years, Janus and Epimetheus meet and exchange orbits. The outer, slower satellite moves into the inner orbit and the inner, slightly faster satellite moves into the outer orbit.

obscures the light from the first body, as seen from the third. In a lunar eclipse, Earth passes between the Sun and the Moon, with Earth casting its shadow onto the Moon. In a solar eclipse, the Moon passes between Earth and the Sun, with the Moon casting its shadow onto Earth's surface.

ecliptic plane The plane of Earth's orbit, thought of as extending out to reach the celestial sphere.

ejecta Material thrown outward during the formation of an impact crater.

elliptical orbit An orbit of one object around another, the shape of which is an ellipse. That planetary orbits can be described as ellipses was first realized by Johannes Kepler.

Elysium Mons A volcano on the Elysium region of Mars. Its summit stands 5.6 miles (9 km) above the surrounding plain.

Enceladus An icy satellite of Saturn that was discovered by William Herschel in 1789. It has the highest reflectivity (albedo) of any known planetary body.

Epimetheus A small satellite of Saturn that is a co-orbital with Janus. *See* **co-orbital satellites** above.

equator An imaginary great circle on a body, such as a planet or the Sun. It divides the body into northern and southern hemispheres and is at every point equally distant from the two rotational poles.

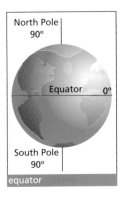

equator

equinox Either of two times each year when the Sun is at the inter- section of the ecliptic plane and the celestial equator, and day and night are equal in length everywhere. The autumnal equinox oc- curs about September 22 and the vernal equinox occurs about March 21.

erosion The action or process by which the surface of a planet, moon, or other object is worn away by the ac- tion of wind, water, ice, thermal or chemi- cal activity, or the im- pacts of meteorites.

eruptive prominence *See* **prominence.**

escape velocity The speed an object must attain to completely overcome the gravita- tional attraction of a body. For Earth, escape velocity is about 25,000 miles per hour (40,300 km per hour).

Europa A Galilean satellite of Jupiter . Eu- ropa has the smoothest known surface in the solar system. It is thought to have a thick crust of water ice.

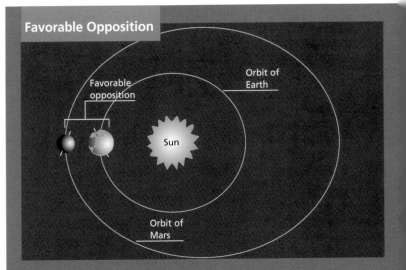

Favorable opposition is the configuration of Earth and Mars at times when Mars is at its closest position to the Sun and about 35 million miles (56 mil- lion km) from Earth, and Earth is between the Sun and Mars, leaving the face of Mars that is turned to- ward Earth fully illumi- nated and easy to observe. Favorable op- position occurs every 17 years around the month of August. The next favorable opposi- tion will take place in the year 2003.

European Space Agency (ESA) A space organization with representatives from Austria, Belgium, Denmark, France, Germany, Ireland, Italy, the Netherlands, Norway, Spain, Sweden, Switzerland, and the United Kingdom. Beginning in 1975, European countries joined forces to develop their own launch programs and now launch commercial communications and weather satellites. ESA and NASA have often worked collaboratively on several projects, including the Hubble Space Telescope.

F

fault A fracture in the crust of a planet or moon in which one side moves relative to the other side, resulting in displacement.

favorable opposition *See* page 215.

flare *See* **solar flare**.

fusion The union of nuclei of certain atoms in a reaction to produce the nuclei of heavier atoms. When the nuclei of certain elements unite, massive amounts of energy are released. Fusion reactions generate enormous energy in the center of a star such as the Sun.

G

galaxy A vast assemblage of stars and associated matter held together by mutual gravitation. Galaxies are typically separated from one another by millions of light-years. *See* **Milky Way**.

Galilean satellite Any of the four largest moons of Jupiter—Io, Europa, Ganymede, and Callisto—discovered by Galileo in 1610.

Galileo (full name Galileo Galilei) (1564–1642) An Italian astronomer and physicist, the first to use a telescope to study the heavens. Among his discoveries were sunspots and their motion across the Sun's surface; the phases of Venus; and the four largest moons of Jupiter.

Galileo probe A probe launched in 1989 to arrive at Jupiter in 1995. The main part of the probe is planned to orbit the planet for two years, gathering data and transmitting images of Jupiter and its satellites back to Earth. A small unit dropped by parachute is planned to relay to Earth information about Jupiter's atmosphere.

Galileo probe

Galileo Regio An area 1,990 miles (3,200 km) across on Jupiter's moon Ganymede. It may be the remains of an ancient impact basin.

Galle, Johann Gottfried (1812–1910) A German astronomer who was the first person to observe the planet Neptune, on September 23, 1846.

gamma rays Extremely energetic photons—electromagnetic radiation of the highest frequency and shortest wavelength. Gamma rays are produced in the cores of stars as a by-product of nuclear fusion.

Ganymede A Galilean satellite of Jupiter, the largest satellite in the solar system. With a diameter of 3,269 miles (5,260 km), Ganymede is larger than the planet Mercury.

gas giant Any of the giant planets Jupiter, Saturn, Uranus, and Neptune.

graben A narrow linear valley that is the result of the faulting and dropping down of segments of the crust

graben

of a terrestrial planet or a satellite.

granule Any of the brilliant spots or cells that mottle the photosphere of the Sun and are caused by the convection of rising hot gas.

gravity A property of matter by which two objects or bodies exert

a force of attraction on each other. In the case of a massive body such as a star or planet, gravity causes a less massive body to fall toward the star's or planet's center and creates a sense of weight.

Great Dark Spot An apparent large storm system on the planet Neptune, images of which were sent back to Earth by *Voyager 2* in 1989. By 1994, images of Neptune from the Hubble Space Telescope indicated that the Great Dark Spot had vanished.

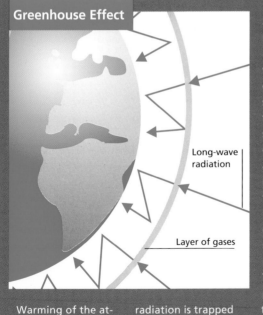

Greenhouse Effect

Long-wave radiation

Layer of gases

Warming of the atmosphere and surface of a planet. Infrared radiation is trapped by certain gases, such as carbon dioxide, and clouds in the planet's atmosphere, leading to gradual warming. Methane, a gas that readily absorbs infrared radiation, comes from such sources as organic decay, arctic tundra, and rice paddies. Carbon dioxide is less effective than methane in absorbing such radiation, but it is far more prevalent in Earth's atmosphere. The surface of the planet Venus is extremely hot, due to a runaway greenhouse effect.

Great Red Spot A reddish, hurricane-like storm system of swirling gases, about as large as Earth in area, that has been observed in Jupiter's southern hemisphere for more than 300 years.

greenhouse effect *See* page 217.

Hellas An impact crater in the southern hemisphere of Mars. It is 1,240 miles to about 1,200 (2,000 km) across—as large as the Caribbean Sea on Earth.

Herschel, Sir William (1738–1822) An English astronomer who discovered the planet Uranus on March 13, 1781.

hot spot A place where magma from Earth's mantle rises into the crust, forming a volcano or volcanic plains. The Hawaiian Islands have been formed as the Pacific plate has passed over a hot spot.

Hubble, Edwin Powell (1889–1953) An American astronomer after whom the Hubble

Space Telescope is named.

Hubble Space Telescope

Hubble Space Telescope (HST) A telescope orbiting Earth. It is designed to observe galaxies in the distant regions of the universe as well as objects in the solar system, in visible, infrared, and ultraviolet light.

Huygens, Christiaan (1629–1695) A Dutch astronomer, mathematician, and physicist who discovered Saturn's moon Titan in 1655 and determined the true shape of Saturn's rings in 1659.

Huygens probe A probe that will be released on a data-gathering descent into the atmosphere of Saturn's moon Titan during the Cassini mission.

Iapetus A satellite of Saturn with one hemi-

sphere that is coated with a dark, carbon-bearing material.

ice volcanism A process apparently shaping the surface of Neptune's moon Triton. Pockets of liquid and gaseous nitrogen trapped under Titan's surface become heated and, under great pressure, break through the surface, flooding and smoothing the surrounding areas before refreezing. The same process produces the geyser-like plumes of nitrogen gas that were observed rising above Triton's surface during the *Voyager 2* flyby in 1989.

impact basin A very large crater often marked by concentric rings or mountains. Examples are the Caloris Basin on Mercury or the various basins on Earth's Moon. Such basins were formed by the impacts of large asteroids or comets.

impact crater A circular depression formed on the surface of a planet or moon by the impact of a meteoroid or comet.

infrared radiation Electromagnetic radiation whose wavelength is longer than that of visible light.

inner planets The planets Mercury, Venus, Earth, and Mars, all of which have orbits that are inside the orbits of the asteroids in the asteroid belt. *See* **outer planets.**

interplanetary Moving or occurring in space between the planets of the solar system.

interstellar Located, occurring, or moving in space between stars, especially the space within the Milky Way galaxy.

Io A Galilean satellite of Jupiter. Io has a diameter of 2,256 miles (3,630 km), about the same as Earth's Moon, and is the most volcanically active object in the solar system.

Ishtar Terra A large highland plateau in the northern hemisphere of Venus. It is about the size of Australia.

Isidis Basin A large impact basin in Mars's northern hemisphere. Isidis Basin has a diameter of 1,180 miles (1,900 km).

Ithaca Chasma A valley system on Saturn's moon Tethys that extends three-quarters of the way around the satellite's surface.

J
K

Janus and Epimetheus

Janus A small satellite of Saturn that is a co-orbital of Epimetheus.

Kepler, Johannes (1571–1630) A German mathematician and astronomer who formulated the laws of planetary motion using the astronomical observations of Tycho Brahe.

Kuiper Belt A region beyond Pluto's orbit that contains icy bodies left over from the formation of the solar system. These bodies are in orbit around the Sun, and among them are thought to be comets with orbital periods of less than 200 years. The U.S. astronomer Gerard Kuiper proposed the

existence of the Belt in 1951.

L

Lakshmi Planum An extensive plateau making up the western part of Ishtar Terra, on Venus.

Larissa A moon of Neptune discovered orbiting beyond the planet's rings by *Voyager 2.*

Lassel, William (1799–1880) An English astronomer who discovered Neptune's moon Triton on October 10, 1846.

lava Molten rock or magma that reaches the surface of a planet or moon. *See* **basalt.**

Leverrier, Urbain Jean Joseph (1811–1877) A French astronomer whose calculations led to Johann Gottfried Galle's discovery of Neptune in 1846.

light-year A unit of length or distance in astronomy that is equal to the distance light travels in one year: approximately 6 trillion miles (9.5 trillion km). *See* **speed of light.**

lithosphere The rigid, outermost layer, containing the crust, on Earth as well as on the other terrestrial planets and satellites.

Lowell, Percival (1855–1916) An American astronomer known for his studies of Mars and for his mathematical calculations that contributed to the discovery of Pluto.

luminosity The amount of energy in the form of light emitted by a star or other celestial object.

lunar eclipse *See* **eclipse.**

lunar Of or relating to Earth's Moon: *lunar* surface; *lunar* orbit.

M

Magellan A NASA spacecraft that orbited Venus from August 1990 to October 1994. It imaged about 98 percent of the surface of Venus at high resolution, using radar that penetrated the planet's cloud layers. *Magellan* was destroyed during its final descent into

Venus's brutal atmosphere.

magma Molten rock material that forms deep below the surface of a planet or moon. *See* **basalt; lava.**

magnetic field The region of space around an object where magnetic forces exist. The field is represented by a set of lines, or paths, along which the magnetic forces are directed. The Sun, Earth, and some of the other planets have significant magnetic fields.

magnetic pole Either of two points on a planet's surface around which the planet's magnetic field lines intersect the surface and where the magnetic field lines are concentrated.

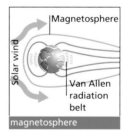

magnetosphere

magnetosphere The region surrounding a planet where the magnetic field of the body is dominant and charged particles from the solar wind become trapped.

mantle The part of the interior of Earth or another terrestrial planet that lies beneath the crust and lithosphere and above the core.

mare Any of various broad, dark, flat areas on the surface of the Moon that are composed of basalt flows. The designation comes from the Latin word for "sea."

maria A plural of **mare.**

mass A measure of the total amount of material that a body contains and that causes it to have weight in a gravitational field.

Maunder minimum A period of 70 years, from 1640 to 1710, when solar activity was very low and there were few sunspots. The period was named after the English astronomer E. Walter Maunder, who noted its characteristics when studying the historical records in the 1890s.

Maxwell Montes A mountain range in the eastern part of Ishtar Terra, on Venus. Named after the Scottish physicist James Clerk Maxwell,

Maxwell Montes contains the highest mountains on Venus, with some peaks reaching heights of nearly 7 miles (11 km).

meteor A meteoroid that passes rapidly through Earth's atmosphere, heating the air along its path and producing a streak of light in the sky. Meteors are also called shooting stars or falling stars.

meteorite A metallic or stony object that has landed on the surface of Earth from outer space. It is the part of a meteoroid that survives passage through Earth's atmosphere.

meteoroid A small metallic or rocky body orbiting the Sun. The Earth encounters many meteoroids but most burn up in its atmosphere before they can reach the planet's surface.

Milky Way or **Milky Way galaxy** The spiral galaxy that contains the solar system. From Earth, the Milky Way appears overhead in the night sky as a luminous band of stars.

Milky Way galaxy

Mimas A satellite of Saturn notable for having a disproportionately large impact crater that

Mariner

A NASA program of unpiloted planetary probes. Six successful missions were launched from July 1962 to November 1973, exploring Mars, Venus, Mercury, and the interplanetary environment. *Mariner 9* was put into orbit around Mars in 1971, becoming the first spacecraft to achieve an orbit around a planet other than Earth. *Mariner 10* was sent to Mercury and Venus in 1973–1975. Images from the probe provided information about the Venusian atmosphere and revealed such surface features as cliffs and craters on the planet Mercury.

is nearly one-third the satellite's diameter.

Mimas

minor planet *See* **asteroid.**

Miranda A satellite of Uranus that is the smallest of the planet's satellites visible from Earth. Among its many striking features is a cliff over nine miles (15 km) high.

molten Melted or liquefied by heat. Magma is *molten* rock, and lava is *molten* before it reaches the surface and cools and hardens.

mons The Latin word for "mountain."

montes The plural of *mons.*

Moon Earth's only known natural satellite.

moon A natural satellite of a planet.

moonrise The first appearance of the Moon above Earth's horizon.

NASA (National Aeronautics and Space Administration) An agency of the U.S. government established by President Eisenhower in 1958 to implement space policy and direct efforts toward the exploration and commercial uses of space.

neap tide A tide of Earth's oceans that occurs twice a month, at the Moon's first and third quarters, when the difference between high and low tide is least. *See* **spring tide.**

Nereid A small satellite of Neptune that was discovered and photographed from Earth in 1949.

northern lights *See* **aurora borealis.**

nuclear fusion *See* **fusion.**

Oberon A satellite of Uranus that is distinguished by its heavily cratered surface and by craters with bright rays of ejecta.

occultation An event in which, from an observer's point of view, an object such as a planet or moon passes in front of a body such as a star or planet. An eclipse is a form of occultation.

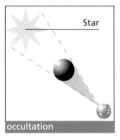

Star

occultation

Olympus Mons A shield volcano on Mars. With a height of 16.2 miles (26 km), Olympus Mons is by far the largest volcano in the solar system.

Oort Cloud A spherical envelope of more than one trillion comets that are believed to orbit the Sun very slowly at a distance of 20,000 to 100,000 AU. The Cloud's existence was proposed in 1950 by the Dutch astronomer Jan Oort.

optical light *See* **visible light.**

orbit The path described by one body in revolution around another body, such as the path of Earth around the Sun or the Moon around Earth.

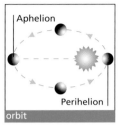
orbit

orbital period The length of time required for one body to complete a single orbit around another body.

outer planets Any of the planets Jupiter, Saturn, Uranus, Neptune, and Pluto, all of which orbit the Sun beyond the asteroid belt. *See* **inner planets.**

ozone A form of oxygen that, in the upper atmosphere of Earth, acts as a barrier shielding Earth's surface from the ultraviolet radiation from the Sun.

**P
Q**

Pandora A small satellite of Saturn that appears to serve as a shepherd for one side of a planetary ring, with Prometheus as shepherd for the other side.

partial eclipse An eclipse of one celestial

body by another such that only a portion of the disk of the eclipsed body is obscured from the observer. *See* **annular eclipse; eclipse; total eclipse.**

partial eclipse

penumbra The region of partial shadow that surrounds the dark central shadow cast by a planet or moon. It is in the penumbra that a partial eclipse is observed.

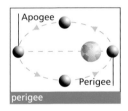
perigee

perigee The point nearest Earth in the orbit of the Moon or of an artificial satellite that is in orbit around Earth. *See* **apogee.**

perihelion The point nearest the Sun in the orbit of a planet, comet, or asteroid. *See* **aphelion.**

permafrost A permanently frozen aggregate of ice and soil, as that found in tundras on Earth and possibly in colder regions on Mars.

photosphere The luminous visible surface of the Sun.

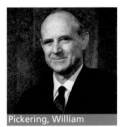
Pickering, William

Pickering, William (1858–1938) An American astronomer who in 1919 predicted the existence of a ninth planet, thus contributing to the eventual discovery of Pluto in 1930.

Pioneer missions *See* page 224.

plana The plural of planum.

planet Any of the nine large bodies, some with satellites, that orbit the Sun in the solar system. Also, any similar body that may be found orbiting a star other than the Sun.

planitia A Latin word indicating "plain"; used in naming low plains or

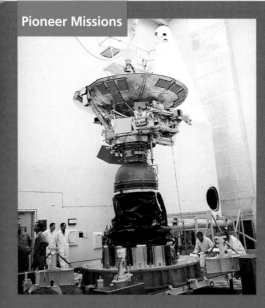

Pioneer Missions

A NASA program of unpiloted planetary probes. Thirteen missions were launched from October 1958 to August 1978. The first three missions were lunar probes that failed to reach the moon but provided important data on the environment of space. *Pioneers 4* through *9* orbited the Sun and studied the inner solar system. *Pioneer 10* flew close to Jupiter; *Pioneer 11* reached Saturn and sent images and new information about the planet back to Earth. The last mission, *Pioneer-Venus*, made the first topographical survey of the Venusian surface.

large regions of lowlands on planets other than Earth.

planitiae The plural of **planitia**.

planum A Latin word meaning "level" or "flat"; used in naming high plains or plateaus on Venus and Mars.

plate Any of the dozen or more slowly moving segments into which Earth's lithosphere is divided. These lithospheric plates float upon and travel over the underlying mantle.

plate tectonics The study of the formation, movement, and interaction of Earth's lithospheric plates. The theory holds that processes deep within Earth drive the movements of the plates, and that stresses at the boundaries of the plates cause much of the volcanic, tectonic, and earthquake activity on the planet.

Pluto Fast Flyby A small probe planned for launch as early as 1998 to fly past Pluto and Charon and investigate

their physical characteristics.

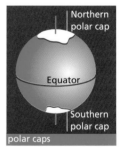

polar caps

polar cap The area surrounding the rotational poles of some planets and moons. It is marked by either permanent or seasonally frozen water or frozen gases. The caps on Mars contain, for example, a mixture of frozen carbon dioxide and water.

pole *See* **magnetic pole; rotational pole.**

Prometheus A small satellite of Saturn that appears to serve as a shepherd for one side of a planetary ring, with Pandora as shepherd for the other side.

prominence A cloud of ionized gas propelled into the Sun's chromosphere or corona by magnetic forces. Prominences occur in a variety of shapes and sizes, and are connected with solar flares or sunspots. A prominence that breaks free from the Sun to surge into space is known as an *eruptive prominence.*

Proteus One of the six satellites of Neptune newly discovered by *Voyager 2.* Proteus is the second largest satellite of Neptune, although its diameter is only 250 miles (400 km).

Proxima Centauri A star that is the Sun's nearest neighbor in the Milky Way galaxy. It is approximately 4 light-years away.

radiation A mechanism of energy transport whereby energy is transmitted through a vacuum. Radiation also refers to electromagnetic energy that includes gamma rays, x-rays, ultraviolet light, visible light, infrared light, microwaves, and radio waves. It also refers to high-energy charged particles, such as protons, electrons, and ions, that are produced by the disintegration of radioactive nuclei or are naturally present in space in the form of solar wind particles and cosmic rays.

radiation belt A band of charged particles that are trapped in the magnetic field of a planet.

radiative diffusion A mechanism by which energy from the core of the Sun moves outward to the surface by a series of absorptions and emissions.

radio telescope A combination radio-receiver and antenna used for astronomical observation by means of detecting radio

waves from outside Earth's atmosphere.

radio telescope

red giant star A reddish star of large volume and luminosity and low surface temperature, with a diameter from 10 to 100 times that of the present-day Sun. Late in its evolution the Sun will expand to become a red giant.

regio A Latin word meaning "region" or "territory"; used in naming large regions of moderate topography, especially on Venus.

regiones The plural of regio.

regolith A layer of dust and rock debris that is found nearly everywhere on the surface of the Moon and other solid bodies of the solar system, such as asteroids. This layer is the result of the pulverizing of the surface by impacts of large and small meteoroids.

revolution The time that it takes for a celestial body to make one complete trip around its orbit, or the action of completing one trip.

Rhea The second-largest satellite of Saturn. The surface of Rhea is very heavily cratered and is thought to be made up primarily of water ice.

ring An apparently flat band of rocky or icy fragments or particles revolving around a planet, such as Jupiter, Saturn, Uranus, or Neptune. These rings are usually found inside of the planet's Roche limit, and always lie in the same plane as the planet's equator.

ring system The complex arrangement of rings and sometimes of satellites around a planet such as Saturn or Uranus.

Roche limit *See* below.

rotation The time that it takes for a celestial body to make a complete turn on its axis, or the action of complet-ing such a turn. Earth takes 24 hours to make one complete rotation.

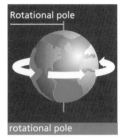
Rotational pole

rotational pole

rotational pole Either of the two points on a body, such as Earth, marking the ends of the axis about which the body rotates.

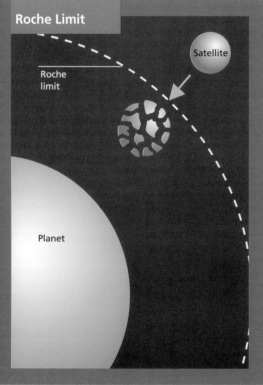

Roche Limit

Roche limit

Satellite

Planet

The distance inside which a planet's tidal forces would shatter an orbiting moon. In the case of Saturn, the Roche limit is about 45,500 miles (73,200 km) above the top of the planet's atmosphere. The inner rings lie within the Roche limit, where Saturn's tidal forces keep the particles from coalescing into larger bodies. The Roche limit was named after the French mathematician and astronomer Edouard Albert Roche (1820–1883).

satellite A relatively small body, such as a moon, that orbits a larger body, such as a planet; also, an artificial device, as a communications or weather satellite, that orbits Earth or another body.

scarp A cliff typically formed by faulting or erosion.

Schiaparelli, Giovanni (1835–1910) An Italian astronomer who observed markings on the surface of Mars that he called *canali. See* **canal.**

sedimentation The process of depositing material, such as sand, on a surface in layers or other configurations.

shepherd satellite—saturn

shepherd satellite A small moon that helps maintain the edge of a planetary ring by gravitational influence. Shepherd satellites sometimes

orbit in tandem, one on each edge of a ring.

shield volcano A volcano formed by the buildup of basalt flows. Successive flows build a volcano with gentle slopes and a broad, dome-like shape. The volcano's base may cover an extensive area.

Shoemaker-Levy 9 comet A comet that broke into 21 fragments before impacting the planet Jupiter over the period of nearly a week in July 1994.

sidereal Measured in relation to a very distant star or stars: a *sidereal* month; a *sidereal* day.

silicate Any of a large number of minerals that contain silicon and oxygen. The bulk of Earth's crust consists of rocks containing silicate minerals .

solar eclipse *See* **eclipse.**

solar flare A violent, temporary release of magnetic energy from the Sun's photosphere. A solar flare sends large amounts of radiation and charged particles into space. As a result of

a solar flare, increased numbers of particles in the solar wind can interrupt radio transmissions on Earth and cause an intensification in the aurora borealis.

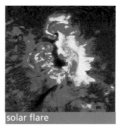
solar flare

solar system The Sun together with the nine planets and all other objects gravitationally bound to it. Also, any similar system centering on a star other than the Sun.

solar wind The flow of high-speed charged particles, such as electrons, protons, and ions, radiating outward from the Sun throughout the solar system. The wind reaches well beyond the planet Pluto and moves as fast as 435 miles (700 km) per second.

solstice Either of two times during the year when the Sun is at its greatest distance from the celestial equator. In the Northern Hemisphere the *summer solstice* occurs about June

22, the longest day of the year and first day of summer; the *winter solstice* occurs about December 22, the shortest day of the year and the first day of winter.

South Pole Aitken Basin An extensive basin on the hidden side of the Moon, about 1,550 miles (2,500 km) across and 7.5 miles (12 km) deep.

space probe A device sent forth from Earth to study planets and other bodies as well as interplanetary space. Probes collect images and data and conduct scientific experiments, sending the results back to Earth. *See* **Galileo probe; Huygens probe; Mariner; Pluto Fast Flyby.**

speed of light The speed at which light travels in a vacuum, about 186,000 miles (300,000 km) per second. *See* **light-year.**

spring tide Any unusually high or low tide of Earth's oceans that occurs at the time of a new Moon or full Moon, when Earth, the Moon, and the Sun are in approximate alignment. *See* **neap tide.**

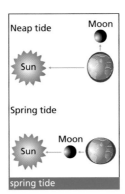

spring tide

star A massive, brilliantly glowing sphere of gas whose energy is generated by nuclear fusion. The Sun is a star of average size.

stratovolcano A volcano having the shape of a cone with steep sides made up of layers of lava and ash.

subduction The descent of one of Earth's lithospheric plates beneath another.

summer solstice *See* **solstice.**

sunspot A dark patch in the Sun's photosphere that is about 1730°C (3150°F) cooler than the surrounding gases. It is caused by a strong magnetic field that blocks convection in the hot gases, thus allowing the region to cool. A sunspot has a dark appearance when viewed through a telescope.

supercontinent A very large continent in the geologic past. The most notable is Pangaea, from which the smaller continents are thought to have split away, starting about 180 million years ago, and drifted to their present positions on the Earth's surface.

synodic Measured in relation to the Sun: a *synodic* month.

T

tectonics *See* **plate tectonics.**

terra A Latin word meaning "land," "earth;" used in naming extensive land masses on Venus and Mars.

terrestrial planet Any of the four inner planets Mercury, Venus, Earth, and Mars that have similar physical characteristics, such as size and average density. The word terrestrial comes from *terra*, the Latin word for "Earth."

tessera A region of highly deformed terrain on Venus, probably formed by tectonic activity. The plural of *tessera* is *tesserae.*

Tethys A satellite of Saturn that has a huge valley system extending nearly around its circumference.

Tharsis Rise A large volcanic area on Mars that is 5,000 miles (8,000 km) across and 6 miles (10 km) high.

tidal bulge A distortion or elongation in the shape of an object, such as a moon, that is caused by the differential gravitational pull of another body, such as a planet.

tidal force A differential gravitational pull that one object exerts on another, affecting the second object's shape or physical integrity. This occurs because of the difference in the gravitational pull, which the first object exerts at different points on the second object.

Titan A moon of Saturn that is the second-largest satellite in the solar system after Jupiter's Ganymede. Titan has an atmosphere that is denser than that of Earth.

Titania The largest of the moons of Uranus.

Titania has a diameter of 1,001 miles (1,610 km).

Tombaugh, Clyde William (1906–) An American astronomer who discovered the planet Pluto in 1930.

total eclipse An eclipse of one body by another such that the disk of the eclipsed body is completely obscured from the observer. *See* **annular eclipse; eclipse; partial eclipse.**

transit The passage of a smaller body, such as the planet Mercury or Venus, across the visible face of a larger body, such as the Sun.

Triton The largest satellite of the planet Neptune, with a diameter of 1,680 miles (2700 km). It has a very thin atmosphere of nitrogen and methane. With a surface temperature of –409°F (–245°C), Triton's surface is the coldest yet observed in the solar system.

Tunguska Event The devastating explosion of an estimated 100,000-ton comet in

the atmosphere over Siberia in 1908. Trees were flattened for nearly 20 miles (32 km) from the shock waves.

Tycho A relatively young crater on the Moon. It is best known for the long "rays" of material ejected during the impact which formed the crater. These rays extend nearly halfway around the Moon.

ultraviolet light Electromagnetic radiation whose wavelength is shorter than that of visible light.

umbra The dark central shadow cast by a planet or moon. It is in this part of the shadow that a total eclipse is observed.

Umbriel A moon of Uranus that is about the same size as its neighboring moon Ariel.

Valhalla A huge cratered area marked by concentric circles, about 1,860 miles (3,000 km) in diameter, on Jupiter's moon Callisto.

Valles Marineris An extensive and very deep canyon system on Mars.

Venera missions A series of unpiloted probes launched by the Soviet Union between the 1960s and the 1980s to explore Venus. The first successful landing on Venus was by *Venera 7* in 1970.

vernal equinox *See* **equinox.**

Viking Either of two successful unpiloted probes, each consisting of an orbiter and a lander, launched by the United States in 1975 to explore Mars.

visible light Light that is in the region of the electromagnetic spectrum that is perceptible to human vision. Also called *optical light.*

volatile A material that melts at a relatively low temperature.

volcanism The activity by which materials such as molten rock and gases are expelled from the interior of a planet or moon.

volcano An opening in a planet's or moon's crust through which molten rock, ash, and gases are ejected onto the surface. Also, a volcano is the mountain gradually built up around a vent from the accumulation of the ejected materials.

W

weight The force with which an object is attracted to Earth or another body by gravity. The weight of an object

becomes less as it moves farther away from the attracting body, whereas the object's mass remains constant.

white dwarf A star of high temperature, low luminosity, small size, and great density that is about the size of Earth and having the mass of the Sun and thus a much greater density.

winter solstice *See* **solstice.**

year The time it takes for a planet to complete a revolution about the Sun, expressed in Earth years or Earth days. A year on Mercury is 88 Earth days; a year on Pluto is about 248 Earth years.

Atlas of the Terrestrial Planets

The maps in this Atlas show only the terrestrial planets, bodies with Earth-like characteristics and surfaces. These include Mercury, Venus, Earth, and Mars. The Moon is also shown.

The surfaces of the terrestrial planets have much in common and yet are all distinctly different. The highlands of Venus, for example, are similar to the continents on Earth and the lowlands are similar to Earth's ocean basins. The shape and distribution of Venus' landmasses, however, are very different from Earth's. The heavily cratered surfaces of the Moon and Mercury are also very similar. To date, only about 45% of the surface of Mercury has been imaged. Thus, other large impact basins, like the Caloris Basin, may exist on Mercury. The highlands of Mars could easily be mistaken for the heavily cratered highlands of the Moon and Mercury.

Many of the features found on the maps that follow are discussed in this book. White areas on the maps indicate areas that have yet to be imaged. The Earth is shown without its oceans so that it can be accurately compared to the other planets.

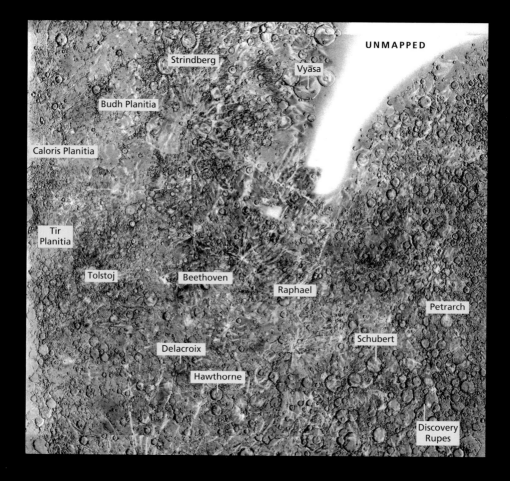

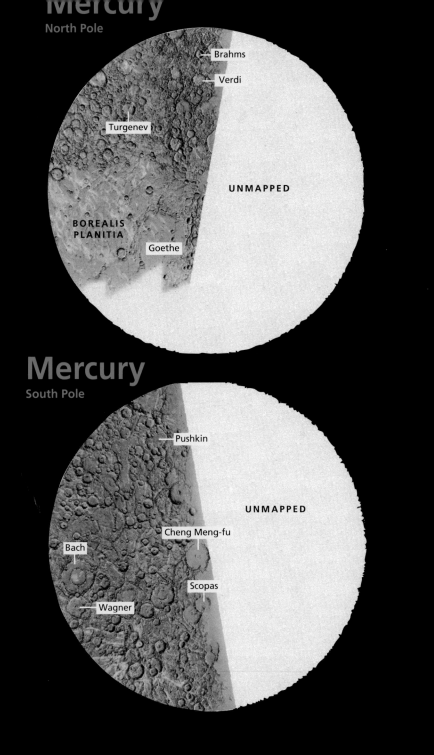

Mercury
North Pole

Brahms

Verdi

Turgenev

UNMAPPED

BOREALIS
PLANITIA

Goethe

Mercury
South Pole

Pushkin

UNMAPPED

Cheng Meng-fu

Bach

Scopas

Wagner

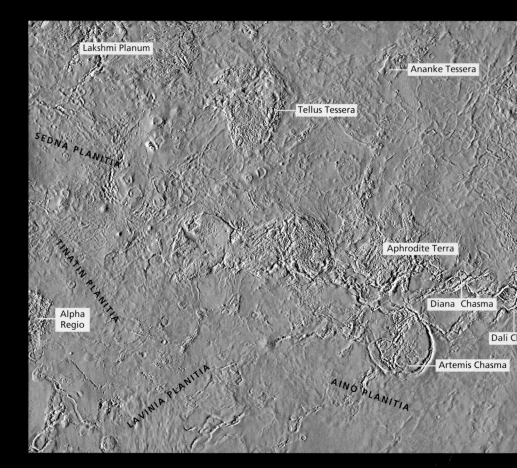

Lakshmi Planum

Ananke Tessera

Tellus Tessera

SEDNA PLANITIA

TINATIN PLANITIA

Aphrodite Terra

Diana Chasma

Alpha
Regio

Dali Cl

Artemis Chasma

LAVINIA PLANITIA

AINO PLANITIA

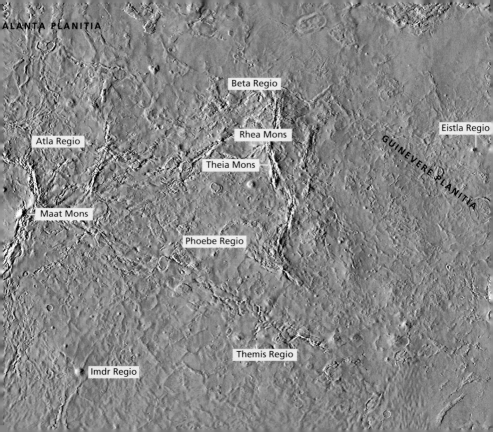

ALANTA PLANITIA

Beta Regio

Eistla Regio

Rhea Mons

Atla Regio

GUINEVERE PLANITIA

Theia Mons

Maat Mons

Phoebe Regio

Themis Regio

Imdr Regio

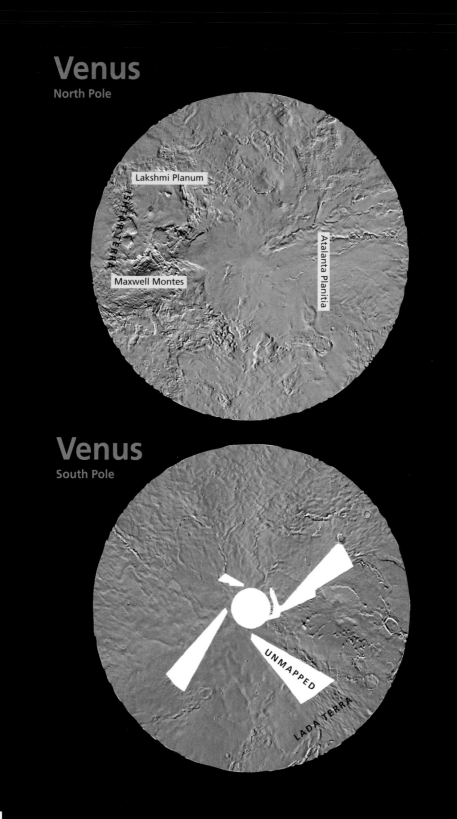

Venus
North Pole

Lakshmi Planum

ISHTAR TERRA

Maxwell Montes

Atalanta Planitia

Venus
South Pole

UNMAPPED

LADA TERRA

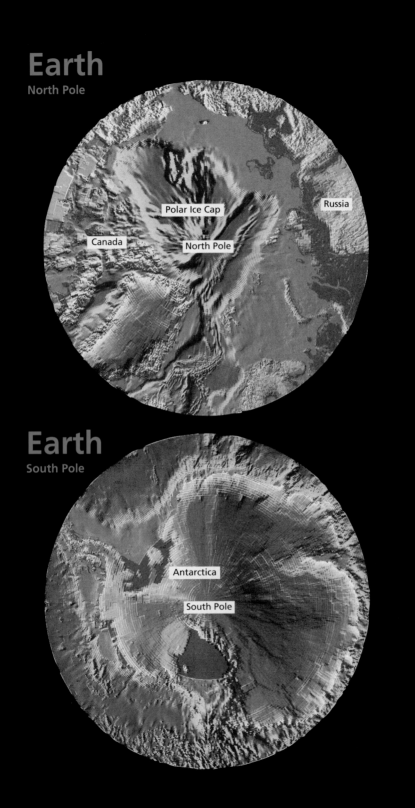

Earth
North Pole

Russia

Polar Ice Cap

Canada

North Pole

Earth
South Pole

Antarctica

South Pole

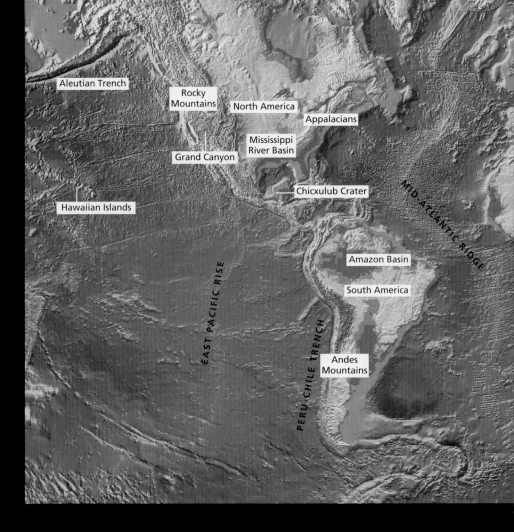

Aleutian Trench

Rocky Mountains

North America

Appalacians

Grand Canyon

Mississippi River Basin

Chicxulub Crater

Hawaiian Islands

MID-ATLANTIC RIDGE

Amazon Basin

South America

EAST PACIFIC RISE

PERU-CHILE TRENCH

Andes Mountains

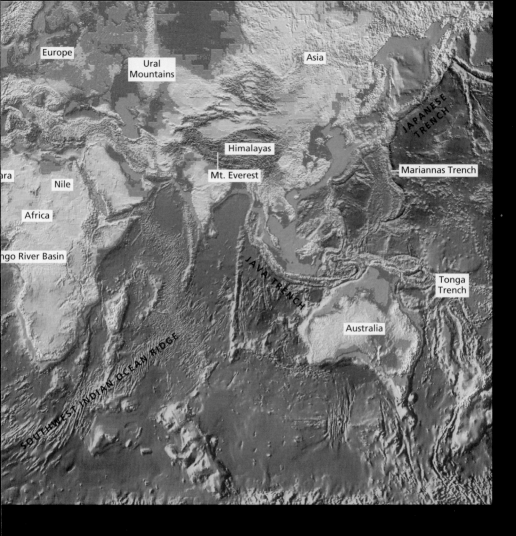

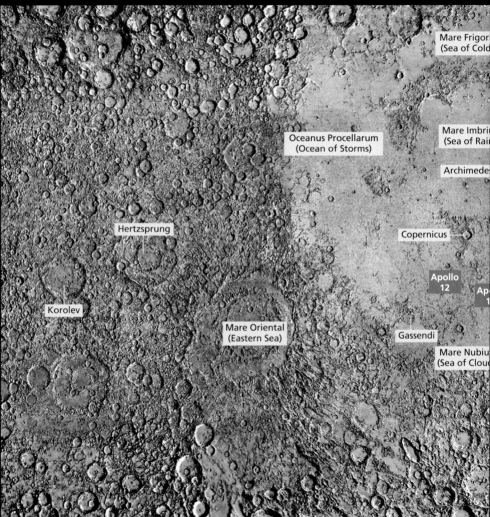

Mare Frigori
(Sea of Cold

Mare Imbri
(Sea of Rai

Oceanus Procellarum
(Ocean of Storms)

Archimede

Copernicus

Hertzsprung

Apollo
12

Ap

Korolev

Mare Oriental
(Eastern Sea)

Gassendi

Mare Nubiu
(Sea of Cloud

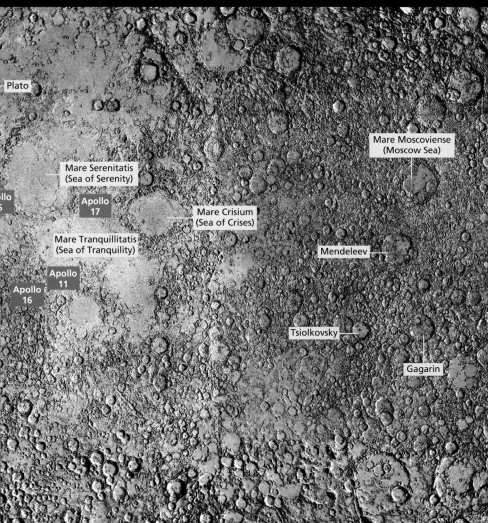

Plato

Mare Moscoviense
(Moscow Sea)

Mare Serenitatis
(Sea of Serenity)

llo

Apollo
17

Mare Crisium
(Sea of Crises)

Mendeleev

Mare Tranquillitatis
(Sea of Tranquility)

Apollo
11

Apollo
16

Tsiolkovsky

Gagarin

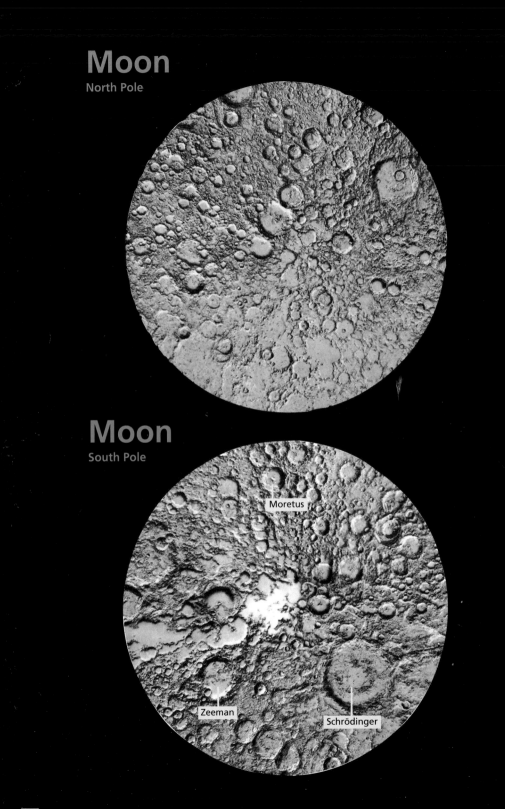

Moon
North Pole

Moon
South Pole

Moretus

Zeeman

Schrödinger

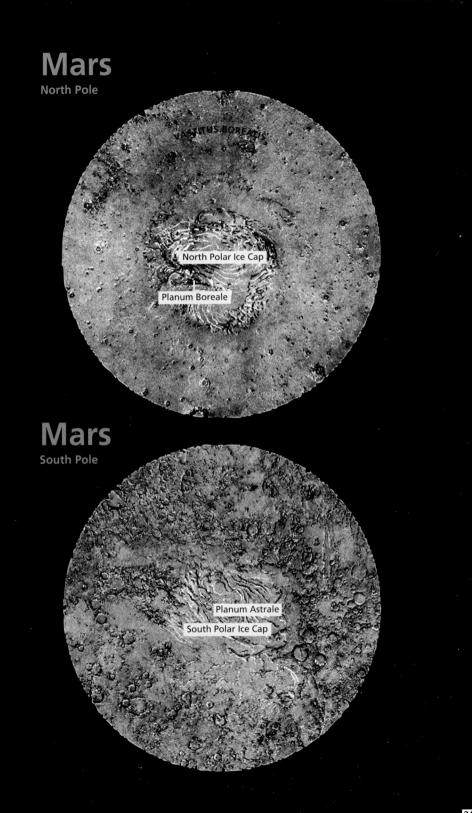

Mars
North Pole

VASTITUS BOREALIS

North Polar Ice Cap

Planum Boreale

Mars
South Pole

Planum Astrale

South Polar Ice Cap

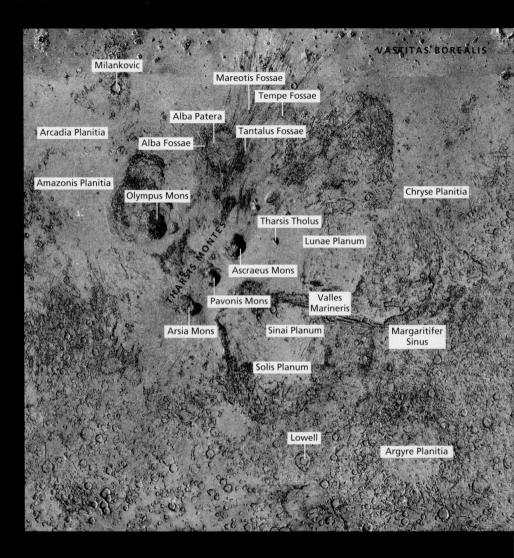

VASTITAS BOREALIS

Milankovic

Mareotis Fossae

Tempe Fossae

Alba Patera

Arcadia Planitia

Tantalus Fossae

Alba Fossae

Amazonis Planitia

Chryse Planitia

Olympus Mons

Tharsis Tholus

THARSIS MONTES

Lunae Planum

Ascraeus Mons

Pavonis Mons

Valles
Marineris

Arsia Mons

Sinai Planum

Margaritifer
Sinus

Solis Planum

Lowell

Argyre Planitia

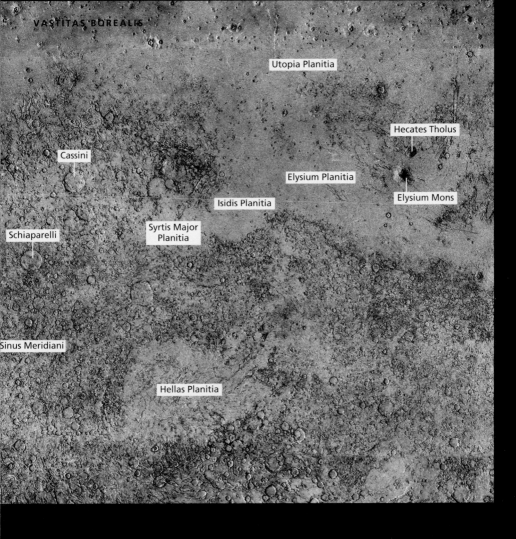

VASTITAS BOREALIS

Utopia Planitia

Hecates Tholus

Cassini

Elysium Planitia

Elysium Mons

Isidis Planitia

Schiaparelli

Syrtis Major Planitia

Sinus Meridiani

Hellas Planitia

Solar System Facts

	Distance from Sun (compared to Earth)	Length of Year (Earth years for one orbit)	Avg. Orbital Velocity (miles per hour)	Eccentricity of Orbit (0 to 1, where 0 is a circle)	Inclination of Orbit to the Ecliptic Plane (in degrees)
Mercury	0.387	0.241	107,127 (172,404 km)	0.206	7.004
Venus	0.723	0.615	78,360 (126,108 km)	0.007	3.394
Earth	1	1	66,638 (107,244 km)	0.017	0.000
Mars	1.524	1.881	53,977 (86,868 km)	0.093	1.850
Jupiter	5.203	11.862	29,214 (47,016 km)	0.048	1.308
Saturn	9.539	29.458	21,564 (34,704 km)	0.056	2.488
Uranus	19.191	84.01	15,234 (24,516 km)	0.046	0.774
Neptune	30.061	164.79	12,147 (19,548 km)	0.010	1.774
Pluto	39.529	248.54	10,603 (17,064 km)	0.248	17.148

	Diameter (compared to Earth)	Mass (compared to Earth)	Avg. or Mean Density (water is equal to 1.0)	Weight of 100-lb Person (shows strength of gravity)	Number of Satellites (over five miles across)
Mercury	0.382	0.055	5.43	38 (17 kg)	0
Venus	0.949	0.815	5.25	91 (41 kg)	0
Earth	1	1	5.52	100 (45 kg)	1
Mars	0.533	0.1074	3.95	38 (17 kg)	2
Jupiter	11.209	317.9	1.33	254 (115 kg)	16
Saturn	9.449	95.18	0.69	108 (49 kg)	18
Uranus	4.007	14.530	1.29	91 (41 kg)	15
Neptune	3.883	17.14	1.64	119 (54 kg)	8
Pluto	0.180	0.002	2.03	5 (2 kg)	1

	Length of day (Earth days for one rotation)	Tilt of Axis (in degrees)	Average Temperature (in degrees)	Atmosphere
Mercury	58.65	2	800°F day (427°C) −300°F night (−183°C)	None
Venus	243.01	177.3	900°F (482°C)	Carbon Dioxide
Earth	1	23.45	59°F (15°C)	Nitrogen, Oxygen
Mars	1.029	25.19	−81°F (−63°C)	Carbon Dioxide
Jupiter	0.411	3.12	−243°F (−153°C)	Hydrogen, Helium
Saturn	0.428	26.73	−301°F (−185°C)	Hydrogen, Helium
Uranus	0.748	97.86	−323°F (−197°C)	Hydrogen, Helium
Neptune	0.802	29.6	−373°F (−225°C)	Hydrogen, Helium
Pluto	6.387	122.46	−419°F (−233°C)	Nitrogen, Methane

Major Satellites

	Planet	Year of Discovery	Diameter Compared to Earth	Atmosphere
Moon	Earth	?	0.272	None
Io	Jupiter	1610	0.285	Sulfur dioxide
Europa	Jupiter	1610	0.246	None
Ganymede	Jupiter	1610	0.413	None
Callisto	Jupiter	1610	0.376	None
Titan	Saturn	1655	0.404	Nitrogen
Triton	Neptune	1846	0.212	Nitrogen, Methane
Charon	Pluto	1978	0.100	Nitrogen

Satellites at a Glance

	Discoverer	Year	Distance from Planet	Orbital Period (in days)	Diameter
Earth					
Moon	?	?	238,860 mi / 384,400 km	27.3	2,160 mi / 3,476 km
Mars					
Phobos	A. Hall	1877	5,830 mi / 9,380 km	0.32	13 mi / 21 km
Deimos	A. Hall	1877	14,580 mi / 23,460 km	1.26	8 mi / 12 km
Jupiter					
Metis	S. Synott	1979	79,510 mi / 127,960 km	0.29	25 mi / 40 km
Adrastea	D. Jewitt, E. Danielson	1979	80,140 mi / 128,980 km	0.30	16 mi / 25 km
Amalthea	E. Barnard	1892	112,660 mi / 181,300 km	0.50	106 mi / 170 km
Thebe	S. Synott	1979	137,880 mi / 221,900 km	0.68	62 mi / 100 km
Io	Galileo	1610	261,970 mi / 421,600 km	1.77	2,256 mi / 3,630 km
Europa	Galileo	1610	416,880 mi / 670,900 km	3.55	1,951 mi / 3,140 km
Ganymede	Galileo	1610	664,870 mi / 1,070,000 km	7.16	3,268 mi / 5,260 km
Callisto	Galileo	1610	1,170,000 mi / 1,883,000 km	16.69	2,983 mi / 4,800 km
Leda	C. Kowal	1974	6,893,500 mi / 11,094,000 km	238.72	9 mi / 15 km
Himalia	C. Perrine	1904	7,133,300 mi / 11,480,000 km	250.57	115 mi / 185 km
Lysithea	S. Nicholson	1938	7,282,500 mi / 11,720,000 km	259.22	22 mi / 35 km
Elara	C. Perrine	1905	7,293,000 mi / 11,737,000 km	259.70	47 mi / 75 km
Anaka	S. Nicholson	1951	13,173,100 mi / 21,200,000 km	631.00	19 mi / 30 km
Carme	S. Nicholson	1938	14,043,000 mi / 22,600,000 km	692.00	30 mi / 49 km
Pasiphae	P. Melotte	1908	14,602,000 mi / 23,500,000 km	735.00	31 mi / 50 km
Sinope	S. Nicholson	1914	14,726,500 mi / 23,700,000 km	758.00	22 mi / 35 km
Saturn					
Pan	M. Showalter	1990	83,260 mi / 134,000 km	0.58	12 mi / 20 km
Atlas	R. Terrile	1980	85,500 mi / 137,640 km	0.60	19 mi / 30 km
Prometheus	S. Collins	1980	86,600 mi / 139,350 km	0.61	62 mi / 100 km
Pandora	S. Collins	1980	88,000 mi / 141,700 km	0.63	56 mi / 90 km
Epimetheus	R. Walker	1966	94,090 mi / 151,420 km	0.69	75 mi / 120 km
Janus	A. Dolfus	1966	94,120 mi / 151,470 km	0.70	118 mi / 190 km
Mimas	W. Herschel	1789	115,280 mi / 185,520 km	0.94	242 mi / 390 km
Enceladus	W. Herschel	1789	147,900 mi / 238,020 km	1.37	311 mi / 500 km
Tethys	G. Cassini	1684	183,000 mi / 294,660 km	1.89	659 mi / 1,060 km
Telesto	B. Smith	1980	183,000 mi / 294,660 km	1.89	16 mi / 25 km
Calypso	B. Smith	1980	183,000 mi / 294,660 km	1.89	16 mi / 25 km
Dione	G. Cassini	1684	234,500 mi / 377,400 km	2.74	696 mi / 1,120 km
Helene	P. Laques, J. Lecacheux	1980	234,500 mi / 377,400 km	2.74	19 mi / 30 km
Rhea	G. Cassini	1672	327,500 mi / 527,040 km	4.52	951 mi / 1,530 km

	Discoverer	Year	Distance from Planet	Orbital Period (in days)	Diameter
Titan	C. Huygens	1655	759,200 mi / 1,221,850 km	15.95	3,200 mi / 5,150 km
Hyperion	W. Bond	1971	920,300 mi / 1,481,000 km	21.28	159 mi / 255 km
Iapetus	G. Cassini	1671	2,212,900 mi / 3,561,300 km	79.33	907 mi / 1,460 km
Phoebe	W. Pickering	1898	8,048,000 mi / 12,952,000 km	550.48	137 mi / 220 km

Uranus

	Discoverer	Year	Distance from Planet	Orbital Period (in days)	Diameter
Cordelia	Voyager 2	1986	30,910 mi / 49,750 km	0.34	16 mi / 25 km
Ophelia	Voyager 2	1986	33,400 mi / 53,760 km	0.38	16 mi / 25 km
Bianca	Voyager 2	1986	36,760 mi / 59,160 km	0.44	28 mi / 45 km
Cressida	Voyager 2	1986	38,400 mi / 61,770 km	0.46	40 mi / 65 km
Desdemona	Voyager 2	1986	38,940 mi / 62,660 km	0.47	37 mi / 60 km
Juliet	Voyager 2	1986	40,000 mi / 64,360 km	0.49	53 mi / 85 km
Portia	Voyager 2	1986	41,070 mi / 66,100 km	0.51	68 mi / 110 km
Rosalind	Voyager 2	1986	43,500 mi / 69,930 km	0.56	37 mi / 60 km
Belinda	Voyager 2	1986	46,800 mi / 75,260 km	0.62	42 mi / 68 km
Puck	Voyager 2	1986	53,440 mi / 86,010 km	0.76	96 mi / 155 km
Miranda	G. Kuiper	1948	80,640 mi / 129,780 km	1.41	301 mi / 485 km
Ariel	W. Lassell	1851	118,800 mi / 191,240 km	2.52	721 mi / 1,160 km
Umbriel	W. Lassell	1851	165,300 mi / 265,970 km	4.14	739 mi / 1,190 km
Titiana	W. Herschel	1787	270,800 mi / 435,840 km	8.71	1,000 mi / 1,610 km
Oberon	W. Herschel	1787	362,000 mi / 582,600 km	13.46	963 mi / 1,550 km

Neptune

	Discoverer	Year	Distance from Planet	Orbital Period (in days)	Diameter
Naiad	Voyager 2	1989	29,800 mi / 48,000 km	0.30	37 mi / 60 km
Thalassa	Voyager 2	1989	31,100 mi / 50,000 km	0.31	50 mi / 80 km
Despina	Voyager 2	1989	32,600 mi / 52,500 km	0.33	93 mi / 150 km
Galatea	Voyager 2	1989	38,500 mi / 62,000 km	0.43	99 mi / 160 km
Larissa	Voyager 2	1989	45,700 mi / 73,600 km	0.55	118 mi / 190 km
Proteus	Voyager 2	1989	73,100 mi / 117,600 km	1.12	261 mi / 420 km
Triton	W. Lassell	1846	220,500 mi / 354,800 km	5.88	1,678 mi / 2,700 km
Nereid	G. Kuiper	1949	3,426,000 mi / 5,513,400 km	365.21	211 mi / 340 km

Pluto

	Discoverer	Year	Distance from Planet	Orbital Period (in days)	Diameter
Charon	J. Christy	1978	12,200 mi / 19,640 km	6.39	789 mi / 1,270 km

Index

ASP—Astronomical Society of the Pacific
CEPS—Center for Earth and Planetary Studies
ESA—European Space Agency
JPL—Jet Propulsion Laboratory
NASA—National Aeronautics and Space Administration
NSO/SP—National Solar Observatory/Sacramento Peak
NRL— Naval Research Laboratory
SPC—Sky Publishing Corporation
SI—Smithsonian Institution
USGS—United States Geological Survey

Photo sources and negative numbers are indicated.

1 PhotoDisc™ Images © 1995 PhotoDisc, Inc.; 2–3 NASA 81-HC-541; 4–5 NASA AS11-44-6610; 6–7 NASA 79-HC-261; 8 NASA, from ASP slide set "The Planetary System"; 9 JPL P-21334; 11 NASA 74-HC-260; 15 © 1995 Roger Ressmeyer—Starlight; 17 Werner Forman Archive/Art Resource; 18 SI 94-12473 (l), SI 94-12476 (r); 19 The Bettmann Archive; 20–21 SI A-56122 (l), Erich Lessing/Art Resource (c), SI 76-11427 (r); 22 SI 94-12471 (t), SI 91-15201 (b); 23 SI 78-17427 (t), Scala/Art Resource (b); 24 Scala/Art Resource (t), SI 72-8748 (b); 25 Scala /Art Resource; 26–27 NASA (underlay), NASA 93-8110 (overlay); 28 JPL P-44131; 29 NASA S80-37633; 30–31 SI 87-2636 photo by Eric Long; 33 PhotoDisc™ Images © 1995 PhotoDisc, Inc.; 34 NSO/SP, from ASP slide set "Secrets of the Sun"; 35 © ESA; 36–37 NASA 74-HC-260; 40 NSO/SP, from ASP slide set "Secrets of the Sun"; 41 NSO/SP, from ASP slide set "Secrets of the Sun", SI 94-12472 (inset); 42–43 NASA 80-HC-219 (l), NSO/SP, from ASP slide set "Secrets of the Sun" (c,r); 44–45 Roland and Marjorie Christen, from SPC slide set "Glorious Eclipses" (tl), Akira Fujii, Hiroyuki Tomioka, and Yonematsu Shiono, from SPC slide set "Glorious Eclipses" (tr), James Curry, from SPC slide set "Glorious Eclipses" (b); 46 NASA 91-HC-378; 47 PhotoDisc™ Images © 1995 PhotoDisc, Inc.; 49 JPL P-39715; 50 M. Robinson/CEPS; 51 CEPS; 52 NASA 73-HC-816 (l), NASA (r); 53 CEPS; 54–55 NASA; 56 JPL P-42388; 57 CEPS/NASA; 59 JPL P-40256; 60–61 JPL P-42392 (l), JPL P-41938 (r); 62 NASA/JPL, from ASP slide set "Venus Unveiled" (l), JPL P-38172 (r); 63 JPL P-38388 (l), JPL P-39175 (r); 64 JPL P-42388; 65 JPL P-37501; 66–67 NASA/JPL, from ASP slide set "Venus Unveiled" (l), NASA 82-HC-40 (c), NASA, from ASP slide set "The Planetary System" (r); 68 NASA AS10-34-5026; 70 NASA S36-151-002 (l), NASA 61B-50-007 (r); 71 NASA AST 13-834; 74 NASA AS17-148-22727; 76 J.D. Griggs, courtesy of USGS (t), V. L. Sharpton, from ASP slide set "Terrestrial Impact Craters" (b); 77 Chris Johns, © Tony Stone Images, Inc.; 78 DLR/German Aerospace Research Establishment (t), Art Wolfe, © Tony Stone Images, Inc. (m), Bob Thomason, © Tony Stone Images, Inc. (b); 79 USGS; 80 NASA 61A-50-057; 81 Don Lowe, © Tony Stone Images, Inc. (l), John Marshall, © Tony Stone Images, Inc. (r); 82 CEPS; 83 NASA STS39-73-044, from ASP slide set "Terrestrial Impact Craters" (l), A. V. Murali, from ASP slide set "Terrestrial Impact Craters" (r); 84–85 USGS; 86 PhotoDisc™ Images © 1995 PhotoDisc, Inc. (l,r); 87 NASA S89-48882 (l), NASA S93-41013 (r); 88 PhotoDisc™ Images © 1995 PhotoDisc, Inc. (l,r); 89 PhotoDisc™ Images © 1995 PhotoDisc, Inc. (l), USGS/NRL (r); 90 NASA 69-HC-901; 91 NASA/GSFC AS11-40-5868; 92 USGS/NRL (tl,tr), NASA/GSFC AS15-88-11967 (b); 93 © Lick Observatory, from ASP slide set "The Planetary System" (l), NASA AS11-40-5878 (r); 94 NASA AS15-1556-M3 (l), NASA AS17-953M (r); 95 © Lick Observatory, from ASP slide set "The Planetary System"; 96 Scala/Art Resource; 98–99 NASA AS11-44-6551 (t), The Bettmann Archive (b), NASA AS15-85-11514 (inset); 100 NASA AS15-86-11603 (l), NASA AS16-119-19171 (r); 101 M. Robinson/CEPS; 102 Akira Fujii, from SPC slide set "Glorious Eclipses"; 104 CEPS/USGS; 105 NASA 77-HC-312; 106 Lowell Observatory Photograph (l,r); 107 USGS/NASA; 108 NASA 77-HC-415 (l), NASA 76-HC-741 (r); 109 NASA, from ASP slide set "The Planetary System"; 110–111 CEPS; 112 Lowell Observatory Photograph; 113 CEPS/USGS; 114 NASA 92-HC-568; 115 USGS (l,r); 116 USGS (t), Stephen P. Meszaros, from ASP slide set "Worlds in Comparison" (b); 117 CEPS/USGS; 118 NASA 93-HC-406 (l), NASA 93-HC-407 (r); 119 NASA S94-45023 (l), NASA S91-52173 (r); 120 M. Robinson/U. Hawaii; 121 M. Robinson/U. Hawaii; 122 The Bettmann Archive (l), CEPS/NASA (r); 124 CEPS/NASA; 125 CEPS/NASA; 127 JPL P-34761; 128 NASA 79-HC-65; 129 NASA 72-HC-42; 131 NASA 79-HC-65; 133 NASA 77-HC-330 (t), NASA 77-HC-474 (b); 134 NASA 79-HC-288; 135 NASA, from ASP slide set "The Planetary System"; 136 CEPS/USGS;

137 CEPS/USGS; 138 NASA 79-HC-286; 139 NASA, from ASP slide set "The Planetary System"; 140 NASA 79-HC-238; 142 NASA 79-HC-296; 143 NASA (t), NASA (b); 144–145 JPL P-23887; 146–147 JPL P-23887 (l), NASA (r); 148 NASA; 149 JPL P-23922; 150 NASA 81-HC-17; 151 JPL P-23927; 152 NASA 81-HC-527 (l), NASA 81-H-589 (r); 154 NASA, from ASP slide set "The Planetary System"; 155 NASA; 156–157 JPL P-23265 (l), NASA 82-HC-76 (c), NASA, from ASP slide set "The Planetary System" (r); 158 JPL P-23113 (l), JPL P-23356 (r); 159 JPL P-23961; 160 NASA 82-HC-78; 161 NASA 81-H-587; 162–163 NASA 86-HC-82; 164 JPL P-29539; 165 NASA 86-HC-82; 166 NASA, from ASP slide set "The Planetary System"; 167 CEPS/NASA; 168 NASA, from ASP slide set "The Planetary System"; 169 JPL P-29520; 170 NASA 89-HC-439; 172 NASA 89-HC-439; 173 JPL P-34632 (tl), NASA, from ASP slide set "The Planetary System" (tr), JPL P-34709 (b); 174 JPL P-34764; 175 NASA, from ASP slide set "The Planetary System" (l), NASA, from ASP slide set "The Planetary System" (r); 176–177 NASA STScI-PR94-17; 178 Lowell Observatory Photograph; 179 Marc Buie, from ASP slide set "The Planetary System"; 183 Betty and Dennis Milon, from ASP slide set "The Planetary System"; 184–185 © ESA (t), National Optical Astronomy Observatories (b); 186–187 Lowell Observatory Photograph, from ASP slide set "The Planetary System"; 188 © ESA; 189 © ESA (l), © Max Planck Institut für Aeronomie, from ASP slide set "The Planetary System" (r); 190 NASA 95-HC-23; 191 Chris Hildreth, Cornell University (l), National Astronomy and Ionosphere Center (r); 195 JPL P-23209; 197 NASA 93-8110; 202 NASA 79-HC-250 (tl), JPL P-24067 (tc), JPL P-34764 (tr), NASA 69-HC-901 (bl), NASA 79-HC-83 (bc), JPL P-29522 (br); 204 NASA, from ASP slide set "The Planetary System" (t,mr), NASA 79-HC-296 (ml), NASA 81-HC-17 (b); 205 CEPS (t,m), CEPS/NASA (b); 206 CEPS (lt,rb), NASA, from ASP slide set "The Planetary System" (lb), JPL P-38170 (rt); 207 NASA 61A-50-057 (t), USGS (bl), JPL P-39715 (br); 208 CEPS (all); Glossary: **Argyre basin** NASA, from ASP slide set "The Planetary System"; **Ariel** NASA S86-27012; **asteroid** JPL P-40449; **canal** Lowell Observatory Photograph; **corona** (2) Kazuo Shiota, from SPC slide set "Glorious Eclipses"; **Galileo probe** NASA S89-48714; **graben** CEPS/NASA; **Hubble Space Telescope** NASA STS061-90-028; **Janus** NASA, from ASP slide set "The Planetary System"; **Mariner** JPL P-12035AC; **Milky Way Galaxy** NASA S90-38110; **Mimas** NASA S80-41849; **partial eclipse** Daniel Good, from SPC slide set "Glorious Eclipses"; **Pioneer mission** NASA S75-23518; **Pickering** JPL P-5347; **radio telescope** © ESA; **shepherd satellite** NASA, from ASP slide set "The Planetary System"; **solar flare** NSO/SP, from ASP slide set "Secrets of the Sun".

More About Planets

Beatty, J. Kelly, and Chaikin, Andrew, eds. *The New Solar System*. Cambridge, MA: Sky Publishing Corp. & Cambridge: Cambridge Univ. Press, 1990.

Chartrand, Mark R. *Planets, A Guide to the Solar System*. New York: Golden Press, and Racine, WS: Western Publishing Co., 1990.

Miller, R., and Hartman, W.K. *The Grand Tour, A Traveler's Guide to the Solar System*. New York: Workman Publishing, 1993.

Henbest, N. *The Planets*. London: Penguin Books, 1994.

Contributors

Production	Editorial	Center for Earth and Planetary Studies	U.S. Geological Survey in Flagstaff, AZ
Elizabeth Kun	Jacqueline Bigford	James R. Zimbelman	Special mention to Mark S. Robinson
	Schelte J. Bus	Robert A. Craddock	
	Sue Causey-Foley	Bruce A. Campbell	also Richard Kozak
	Benjamin Chadwick	Andrew K. Johnston	
	Robert Costello	Victoria A. Portway	**NASA/Goddard Space Flight Center**
	Sherri S. Dietrich	Priscilla L. Strain	
	Kristen Holmstrand	Michael J. Tuttle	James B. Garvin
	Elizabeth Mitchell	Karen L. Peters	
	Michael Pistrich	Rosemary E. Steinat	
	Jane Redmont	Donna J. Slattery	
	Hilary Sardella	Ruth A. McGrail	